DIY Comms and Control for Amateur Space

Sandy Antunes

DIY Comms and Control for Amateur Space

by Sandy Antunes

Printed in the United States of America.

Published by Maker Media, Inc., 1160 Battery Street East, Suite 125, San Francisco, CA 94111.

Maker Media books may be purchased for educational, business, or sales promotional use. Online editions are also available for most titles (*http://safaribooksonline.com*). For more information, contact our corporate/institutional sales department: 800-998-9938 or *corporate@oreilly.com*.

Editor: Patrick Di Justo
Production Editor: Melanie Yarbrough
Proofreader: Charles Roumeliotis
Interior Designer: David Futato
Cover Designer: Ellie Volkhausen
Illustrator: Rebecca Demarest

June 2015: First Edition

Revision History for the First Edition

2015-05-27: First Release

See *http://oreilly.com/catalog/errata.csp?isbn=9781449310660* for release details.

978-1-449-31066-0

[LSI]

Contents

Preface

The picosatellite field has exploded, going from less than 10 launched per year (back in 2009-2011, when I did my Calliope build) to over 100 launches *this year* alone. We have gone from requiring a large university or lab to back a project, to a point where even a motivated high school can build one. We are in a Do-It-Yourself (DIY) space age, and I'm proud (maybe even a bit smug) that Calliope has been part of that. In fact, I'm building version 2 of my hardware, using advances designed by others who were, in part, inspired by my Project Calliope!

While waiting for launch, I've written these four short DIY satellite books for *Make:*, each of which has sold several thousand copies. I get emails from DIY builders of CubeSats and TubeSats and PocketQubes asking my advice. These emails often start off with "I was inspired by your project and have a technical question about," followed by a technical question of such complexity that it's clear they are already pushing the picosatellite frontier way past what I did with my little basement-build. There are CubeSats flying this year that include my work as an inspiration.

I was hired by Capitol Technology University to teach college students about satellites and systems engineering and physics. They hired me, to a large degree, because attempting Calliope impressed them. I am full-time faculty, starting my fifth year there, and just got promoted to Associate Professor. Because I haven't launched yet, and I'm helping college students work towards their own satellites, I donated the bulk of my ground station hardware to the not-for-profit Capitol lab I run. This donated hardware is jumpstarting students' engagement in topics like ham radio and payload commanding and stuff. My focus is on what I can do for the next generation of space enthusiasts. In addition, I've become an advocate for the so-called *STEAM* movement, that adds an *A* for Art into the usual STEM curricula.

Life is not a race because the only finishing line is the future, and that is unknown. So why not be the people that create that future? Read, think, build, interact, then repeat. As long as you are moving forward, you are going somewhere.

Conventions Used in This Book

The following typographical conventions are used in this book:

Italic
: Indicates new terms, URLs, email addresses, filenames, and file extensions.

`Constant width`
: Used for program listings, as well as within paragraphs to refer to program elements such as variable or function names, databases, data types, environment variables, statements, and keywords.

`Constant width bold`
: Shows commands or other text that should be typed literally by the user.

`Constant width italic`
: Shows text that should be replaced with user-supplied values or by values determined by context.

This element signifies a tip, suggestion, or general note.

This element indicates a warning or caution.

Safari® Books Online

Safari

Safari Books Online is an on-demand digital library that delivers expert content in both book and video form from the world's leading authors in technology and business.

Technology professionals, software developers, web designers, and business and creative professionals use Safari Books Online as their primary resource for research, problem solving, learning, and certification training.

Safari Books Online offers a range of plans and pricing for enterprise, government, education, and individuals.

Members have access to thousands of books, training videos, and prepublication manuscripts in one fully searchable database from publishers like O'Reilly Media, Prentice Hall Professional, Addison-Wesley Professional, Microsoft Press, Sams, Que, Peachpit Press, Focal Press, Cisco Press, John Wiley & Sons, Syngress, Morgan Kaufmann, IBM Redbooks, Packt, Adobe Press, FT Press, Apress, Manning, New Riders, McGraw-Hill, Jones & Bartlett, Course Technology, and hundreds more. For more information about Safari Books Online, please visit us online.

How to Contact Us

Please address comments and questions concerning this book to the publisher:

> Make:
> 1160 Battery Street East, Suite 125
> San Francisco, CA 94111
> 877-306-6253 (in the United States or Canada)
> 707-639-1355 (international or local)

Make: unites, inspires, informs, and entertains a growing community of resourceful people who undertake amazing projects in their backyards, basements, and garages. Make: celebrates your right to tweak, hack, and bend any technology to your will. The Make: audience continues to be a growing culture and community that believes in bettering ourselves, our environment, our educational system—our entire world. This is much more than an audience, it's a worldwide movement that Make: is leading—we call it the Maker Movement.

For more information about Make:, visit us online:

Make: magazine: *http://makezine.com/magazine/*
Maker Faire: *http://makerfaire.com*
Makezine.com: *http://makezine.com*
Maker Shed: *http://makershed.com/*

We have a web page for this book, where we list errata, examples, and any additional information. You can access this page at: *http://bit.ly/diy-comms-control-amateur-space*.

To comment or ask technical questions about this book, send email to *bookquestions@oreilly.com*.

1/Overview

Talking to Machines

- "A workman is only as good as their tools." [anon]
- "Only a poor workman blames his tools." [anon]
- "A bad worker quarrels with his tools." [anon]
- "Give me a lever and a place to stand, and I will move the world." [Archimedes]

This is a book on how to talk to satellites. Most communications books of this type will welcome you to the exciting world of radio. And I agree. Nearly every ham radio person or AMSAT member or communications enthusiast loves what they do. They revel in the details, and they get great pleasure from pulling off magnificent acts of technical competence to span this world—and others—using simple electronics.

I only want to talk to machines. I just want it to work. I want my satellite communications as a commodity, not an art.

This book, therefore, pays utmost respect to those who are ever advancing the performance frontiers, but tackles it from the base utility view of a plebe. This book will simultaneously provide you with a lay view of how you can talk to satellites, and a peek into the rapidly changing world of satellite radio communications.

Rapidly changing is almost an understatement. In the month before I wrote this introduction, 64 CubeSats were launched or deployed. One new 3000-station comm network was set up by a single *amateur* operation. So this book will, by the time it's published—one month after final edits—probably be ignorant of the newest advances. For this reason, I talk very little on specific models of hardware or protocols, and instead focus on system engineering principals required to assess then develop your entire ground and communications network.

In this age of Making, you can find endless case studies of hardware. This book is intended as the *missing manual* to provide understanding of the entire end-to-end process necessary for accurately commanding and then receiving data from your satellite. If you only walk away with two pieces of knowledge from this book, let them be these: calculate your actual bandwidth and don't overpromise on your mission (which often will deal in kilobytes of data, not megabytes or gigabytes) until you run your comms budgets, and understand that satellite radio is just like "ground radio plus Doppler shift" (Doppler shift is covered in Chapter 3).

Some people view communications as a means to an end; I admit I do. Hopefully this book will convey to you not just the current means, but the trends and directions that are creating new ways of playing with our classical elements of signal, gear, and smarts to let anyone—even me—better talk to machines. In one month in 2013, there were 65 CubeSats launched. What will the future bring—crowding and less spectrum, or shared networks and easier access? This book will hopefully ground you in the core systems engineering principles needed to ensure your satellite has reliable, consistent, accessible communication.

Operations is always a compromise between improvement and risk, between wants and costs. A good *operations mentality* is to understand that risk is not binary, and that "pause and assess" is more important than quick decisions. Early in your design stage, you can go for wacky ideas and embrace risk. Once you are maintaining your system, however, incremental improvement and risk avoidance are more important. In any project, multiple types and viewpoints are not just *good* but essential for strong mission operations. Always try to fall back to:

1. Why are we doing this?
2. How does this serve the mission?
3. Is this the standard approach?
4. Is this the best approach?

Parts of a Comm System

There are several parts to a comms system. To get data down from the satellite, you need a downlink system. This can be as simple as a premade Cubesat transmitter board, driven by your satellite CPU of choice, sending through a simple tape antenna, and being picked up by a ham radio enthusiast using a simple Yagi directional antenna and a handheld radio. "Beep," hears the ham from the satellite. Connect a laptop computer to the radio, and that beep can be decoded to read "Hello world." We call these *beepsats*, though often not favorably—they are old style missions, and current picowork is doing far more with satellites.

Or, you can use a custom transponder driven by your satellite CPU to send data via an aimed directional antenna to a NASA White Sands ground dish capable of receiving tightly focused signals. The data is then processed and routed by their network to your Mission Operations Center, where all the data is sent to different computer consoles based on their function (health data, science data, etc.) for individual handling by your team of skilled mission operators.

Finally, you might have a contract with a comm provider to relay data via their existing network to an Internet center, which allows any distributed user to access it via smartphone or similar device.

Whichever way you get data from your satellite, it's still the same principles. Your satellite data is parsed by the satellite CPU and converted with software and hardware into a radio signal. This is sent by satellite radio hardware out through an antenna and received by, well, whomever is handling your comm receiving. On the ground, hardware and software decodes the data and massages it into a meaningful form, and delivers it to your user or users.

Similarly, to command or control your satellite, you need an uplink system, a way to get data from the user to the satellite. The flow for uplink is to have ground software generate commands that get translated by hardware or software to something the satellite can listen to. These signals get sent by the comm network to the satellite, which receives it via an antenna.

The signal is decoded by hardware and software and then parsed or interpreted by the satellite's computer.

Note that these uplink components are pretty much the identical components as for downlink, just pointing in the other direction. You won't have to invest in two systems, just one. The transceiver both receives and transmits. The antenna on the satellite won't change. You may use one ground antenna (to either send or receive, but not both at the same time), or two antennas—one to send, one to receive. The biggest difference is the level of software required.

Many amateur satellites adopt a mode where anyone can use the data, but only specific designated sites can command the satellite. This is both a security and practicality issue. Much as commercial radio has one producer and many listeners, a good satellite should have only one or two commanders, and many amateur listeners.

In terms of implementation, receiving data and doing something useful with it is handled by one software set, which you can distribute to anyone interested in using your satellite. Creating command loads is a separate bit of software, which usually is not widely provided to the world. Commanding a satellite is significantly more involved than receiving data. Think of the difference between writing and performing a musical piece, versus just listening to it.

Command software packages exist, with no clear lead or winner on which is best. We'll cover several, including CFS, Hummingbird, and home-built AX.25 packet systems.

Encoding and decoding the signal to a format (like the oft-mentioned AX.25) can be done by software—computer code that converts "turn on satellite" into a series of rigidly formatted 8-bit packets. Or by hardware, most often a programmable chip (e.g., a PIC) that does the conversion for you.

The transceiver hardware is fundamental radio gear, attached to your antenna. We'll discuss some configurations and a heap of books from the ARRL that can teach you more.

The satellite receiving antenna and receiver hardware are something you already bought or built. They mimic the ground system in that they must use the same frequency, and have an appropriate power level to connect to each other. We'll discuss how these interact and the concept of *link budgets*, which lets you determine how much data you can transmit in any given time.

You will find your comm system defines the fundamental limits for your mission in terms of the amount of commanding you can do, and the amount of data you can receive. By analogy, if you had a cell phone with a 1GB/day data cap, you're obviously limited to 1GB/day of data. You must also consider whether your phone can see the network (yay, 4 bars!) or is out of contact (no signal, try again later).

Satellites have these limits too. They arise from understanding (a) how fast your radio system can transmit, and (b) how often you can make contact with your satellite. The number of contact *passes* you get with your satellite, multiplied by the number of minutes each pass lasts, multiplied by the data rate your hardware can sustain during a pass, yields your total daily data limits for your mission.

Comm for the Impatient

OK, so you want to quickly build a satellite comm test rig. Here we go, starting with Sputnik-level hardware.

Two Walkie-Talkies

Turn on both walkie-talkies. Tape down the *beep* button on the one representing the satellite. Now have a friend run around the block as the *satellite*, and notice when you can and cannot hear the beep.

The range of these walkie-talkies is limited, so you'll probably find you only hear the beep when your friend is both close and in line of sight (no buildings blocking). It also helps to point the walkie-talkie in their direction.

This illustrates the basics of satellite comm. Everything else is just a fancier way to send more meaningful data than a beep, and to enable you to reply to the satellite. But this exercise does illustrate the core properties we're dealing with in this book.

Flatsat

Now let's build a good test setup for prototyping your future satellite build. The key components are a realistic low-powered *flatsat* or satellite simulator, combined with a laptop-based ground system for doing real data sends and receives. The primary hardware difference between this *bench test* rig and your eventual final station is primarily in the antenna sizes (small) and transceiver power (not many watts) used in this test rig.

Put simply, add a bigger antenna and more watts in the transmitter, and you may be ready to fly with something similar to the "$50 satellite" Eagle-2 (*http://50dollarsat.info*) with handheld radio plus the actual satellite, shown here:

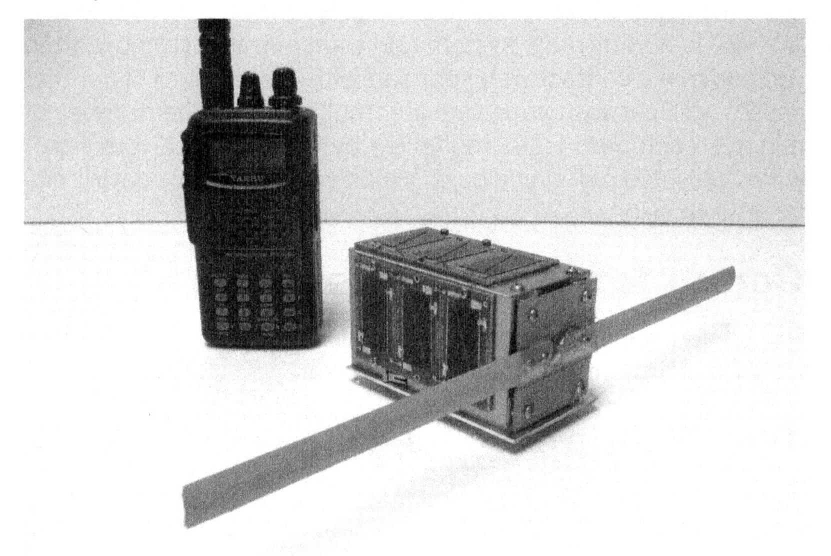

Bidirectional Bench Test Hardware

- Arduino with transmitter and receive modules and a simple wire antenna
- Laptop
- SDR dongle attached to an antenna

The Instructables (*http://bit.ly/1PgmoRL*) on downloading NOAA satellite images with a laptop is a great start. You can also look up *SDR satellites*, short for software-defined radio, and find a wealth of always-current articles on this.

Hook up the modules to the Arduino as per these Instructables or BuildCircuit plans. Let's say you have an Arduino, and a Radiometric transceiver hooked up to the serial lines of the Arduino. Conceptually, a transmitter is identical to a sensor. A sensor (as in Book 3, *DIY Instruments for Amateur Space*) sends signals that the Arduino converts to digital—either through a protocol like I2C or serial lines, or via the built-in ADCs. Similarly, a transmitter involves the Arduino sending out voltages along the pins expected by the transmitter. Most often, this is via serial lines.

Program the Arduino to send a string like *hello world* if and only if the receiver gets the string *go*.

Send the string *go* from the laptop. Wait for the return message.

If this works, you have a basic working comm system.

Power, Range, and Licensing

The first step is to determine what frequencies are available for your task:

Amateur low-power
: What frequencies are legal, and what gear do you have?

Amateur high-power
: Apply for allocation. For satellites, IARU provides guidelines on available frequencies and how to request use of them.

Commercial low-power
: What gear do you have? (The company already got FCC approval for that gear.)

Commercial high-power
: What gear do they require to use their comm network?

Low power means short range. It's the power level that toy store walkie-talkies, hobby shop radio-controlled (R/C) cars, and electronics store WiFi routers use. The power is low so it doesn't interfere or overlap with others in a significant way. The manufacturer of the device has ensured it has passed the tests to comply with (in the US) Federal Communications Commission (FCC) requirements.

In fact, these devices all include a little sticker saying the device follows FCC guidelines. If you modify these—increase the signal or power, change the antenna to improve range—you may actually be violating the licensing by making them more intrusive.

Notice that all high-power comm requires licensing. The FCC coordinates all US licensing, and other countries have their equivalent of the FCC. Both amateur radio and amateur satellites are allowed to use a portion of the radio range for their work, but there are legal and licensing regulations for using that.

In the US, amateur radio enthusiasts can take their ham radio license tests—"Technician Class" is the first level—to talk worldwide using radio. Higher power levels or longer broadcast intervals (such as a radio station) or using radio frequencies not allowed for amateurs requires additional licensing.

"Data" Versus "Packets"

To keep things moving, we'll somewhat interchangeably use *data* and *packets*. Technically, a *data packet* is one element, and many packets together form data. Heck, *data* is already plural—a single piece of data is a *datum*. We transmit radio data packets, which get collected by whatever is hooked to the receiver to form useful data. When we get into technical specs, we will be sending individual packets (which then get assembled into useful data).

For example, packet radio sends a predefined set of individual radio packets, which together can be combined into (for example) a 2D picture. The packets are the data sent, the image is the full data item received. Packets combine to larger data forms. Make sense?

All of this is a fancy way of saying a cheap, short-range walkie-talkie doesn't require a license to use, and the manufacturer got it certified as saying "doesn't need a license." Amateur radio gear requires you take a test, then only use it for specific radio frequencies and according to specific legal requirements (such as always announcing your identity when transmitting). Radio stations and other broadcasts, even at low power, require following specific rules and keeping to specific radio frequencies. The longer the range your radio signals travel, the more you have to deal with licensing issues.

Higher Power Licensing

"High power" equals "long range," so for satellite use, we need high-power solutions. However, you should study the low-power implementations because that is how your test apparatus works. For a given allocated high-power frequency, you can build low-power (unlicensed or commercial) test rigs in your lab for integration and testing. Your high-power licensing is only needed for the *main event*. I keep coming back to *allowed radio frequencies*. Your computer's WiFi router has a specific frequency allocated to it. Your cordless phone has a different frequency, and your smartphone yet another. AM and FM radio stations each have their own little bands of radio frequencies. Meanwhile, broadcast digital TV uses yet another set of frequencies. Airport radar occupies another range of frequencies. The radio spectrum is carved out into different frequency *bands* and who gets to own or use each band is fiercely negotiated between stakeholders—companies, consumers, emergency services, and amateurs.

The result of all the fighting is that some bands are potentially open for satellite (or drone or hobby or ham) communication, depending on how you plan to share it. For low-power or short-term use, things are easier. Higher power requires more advance preparation and potential licensing.

In this model, though, we are talking about how you get permission for whichever device (ground or satellite) that is transmitting. Just receiving data, it turns out, is easy. Just as you need a license to run a radio music station but anyone can listen, you need a license to run a satellite but anyone can listen to your satellite.

Establishing a Ground Network

You can always receive ANY radio signal without a license. Receiving is passive—you build a box that picks up the signal (see "Are You Receiving Me?" on page 11). This means you can use a scanner to listen to police chatter or pirate radio, without fear. More practically, you can (and should!) use your radio gear or laptop plus an SDR to listen to satellites. For example, you can get radio data from NOAA weather satellites to build your own weather maps in real time, before the Weather Channel broadcasts.

Transmitting, however, requires you use a licensed device in the appropriately licensed way. Transmitting means any transmission—from ground to satellite and from satellite to ground are both transmission. In order to send commands to a satellite or receive data from a satellite, you must have both a transmitter and a receiver.

- Send commands: data transmitted from the ground (licensed) is heard by the satellite receiver (unlicensed).
- Receive data: data transmitted from the satellite (licensed) is picked up by a receiver on the ground (unlicensed)

Sending up data from ground to satellite is the uplink. Sending data from the satellite to the ground is the downlink. In both cases, the transmitter must be licensed, but the listener is passive—anything that hears the signal can use it.

Further, for amateur radio, data sent down with IARU (or equivalent) approval must be unencrypted with a data format specified to everyone. However, the data you use for commanding can be private and not provided to anyone.

From this, we get two models. The design model clearly indicates both the satellite and commanding station need licenses, because they are both transmitting. The user model indicates that anyone transmitting commands needs a license, but you can have people help with receiving data and those users don't need a license to help with the downlinks.

From the user model, this means you can build a very large downlink network—to get your satellite's data—among the general public. At the same time, you can legally restrict the number of participants that can actually command your satellite. The uplinkers have to be both licensed (for transmitting at high power) and be privy to your secret command packet format.

Add in that you typically need to send very few command packets, and downlink is your primary communication budget. This is similar to upload/download rates for Earth-based Internet providers: they assume you are taking in a lot of content (streaming videos, web pages, incoming emails, tweets you are following) and sending out proportionally fewer (items you upload to websites, emails you write, tweets you author). For your satellite, you likely have small command uplinks akin to "OK, send me data," "hey, here's a new instrument config," and "here's where I'd like you to point to next," and proportionally larger data downloads: "OK, here's that picture you asked me to take" and "right, here's the 12 hours of data you've been waiting for."

In summary, a good ground network will have lots of sites to receive data and share or forward it to you, and a much smaller group authorized and given the access so they can issue commands to your satellite. Data down is public, and commands can be private, so the model matches pragmatic use as well.

Are You Receiving Me?

Yes, you can legally receive any radio signal transmitted. But also, legally, receiving a coded or encrypting signal than breaking that code can get you into trouble. Receiving it is legal, breaking the code might not be, depending. Just because you can receive the signal from a pay-per-view broadcaster does not give you the right to decrypt it and view it for free. That's a separate issue not dealt with here. This entire book is about signals either broadcast *in the clear* (unencrypted, anyone allowed to listen) or signals that you have a right to decrypt.

Breaking International Law

I have had inquiries along the lines of, "Can I launch a picosatellite so people in [oppressed country X] can communicate despite censorship attempts by their local governments?"

Satellites are by default world-spanning. Any country can launch a satellite that then occupies the airspace of other countries. Any communications satellite can be used to allow disenfranchised people to communicate.

Whether you can do so legally is a different matter, and the answer is usually *no*. If you license via the IARU, you are bound by their laws, which tend to respect the laws of whichever countries you fly over. In practical terms, this means using a satellite to circumvent agreed international radio rules is not recommended.

This is akin to saying that because "X" is legal in your country, you have a right to do "X" in a foreign country. If you wish to use a satellite as a communications station to break the laws of a given country, please consult a specialist in international law. Who knows, your plan might be possible. If a given country doesn't have a law forbidding your action (as advised by your lawyer), you might even have found a new business niche.

If you get arrested, note I said "consult a lawyer," and I claim no responsibility. If you succeed and make millions, though, feel free to buy me lunch and thank for me for my advice.

Data Jargon

Different books and sites use different words to mean the same thing. Both *telemetry* and *data* mean data downlinked from the satellite. "Telemetry" is generally any satellite housekeeping/health and safety information, while *data* is often shorthand for *instrument data* or *science data*. They are often interchangeable: "we are receiving telemetry" and "we are receiving data" might mean the same thing. Or, "we are receiving telemetry" might mean the short health-and-safety info is arriving, while "we are receiving data" means instrument and payload data is arriving. A data archive, for example, usually just distributes the science data, with the housekeeping telemetry folded in as headers or comments.

There are also levels to science data. Level 0 is the raw, unprocessed data. Level 1 has some basic conversions applied, including translating from *instrument numbers* to physical units and calibrations to remove known data artifacts. Higher levels add more post-processing to make the data more useful to end users.

Housekeeping Data

There is key satellite telemetry data you will need to display on your ground system. These fall into the housekeeping (HK) category, items that tell you on the ground the health and safety of your satellite in orbit.

You will usually compare the HK values, on the ground, with the ideal operating conditions.

Housekeeping: Bus and Instrument Are Separate

- Temperatures
- Voltages (bus per system, payload per instrument)
- Power levels and battery status
- Data fill rates
- Spacecraft clock
- Last uplink, last downlink (who accessed and when?), buffers
- Smart instrument status details (calibration levels, etc.)
- Attitude (inertial or other)
- Position (inertial or other)

In addition to displaying the value of these Health and safety items, your design should include an assessment of what the valid or good (green) ranges are, when the values are starting to get worrisome (yellow), and when a value or subsystem is in serious danger (red).

For example, if your power bus is supposed to supply "3.5 volts," what does that really mean? It will not supply 3.5 volts exactly. You need to understand and track the allowable limits. For example, your system might work in the range of 3.3-3.6 volts, but gets flaky below 3.3 volts and above 3.6 volts. At the range of 3.0 volts or lower, or above 4.0 volts, the system may suffer damage and need to power down. So your green limits are 3.3-3.6 volts, your yellow limits are 3.0-3.3 volts or 3.6-4.0, and your red limits are <3.0 volts or >4.0 volts.

Alerts and Pages

A good ground system will take the HK data, compare it against your desired ranges, then alert you if anything is out of spec. This is the equivalent of the idiot lights on your car—*oil low* or *check engine* flash on if a monitored parameter is out of spec, warning you to examine it more deeply.

Displaying these as color bars in your command software is an easy way to, at a glance, assess the status of your spacecraft. Green = good, yellow = keep an eye on it, red = trouble. This intuitive system is used for the big satellites, and should be used for yours.

Limits Displays: Green, Yellow, Red

An alert is your system notifying you something is out-of-spec. A page is when your system actively tries to reach you tell you about it, via a text or a phone page or a tweet or a robot named Alfred the Butler that walks over and taps you on the shoulder to hand you an urgent note. Software typically sends pages by email or text message to a preselected address if any HK data moves into the yellow or red range. Email is the most common way to send pages. Some missions, such as NASA's Tropical Rainforest Measurement Mission, developed an iPhone app to push these pages to a phone.

Data Integrity

Regardless of the power and means for transmitting data, there's a separate issue of data integrity. The core methods to think about include having a verification method, such as checksums, for ensuring your data is complete. Allow for redundant transmission and the ability to retransmit—in essence, expect errors will occur and save some bandwidth to recover from them. In general, assume you have to be able to assemble full data sets from pieces or packets that arrived out of order. As long as each packet is tagged with a timestamp, you can reassemble on the ground in any decent database. Further, use the health and safety data to qualify how good or bad a particular data set is, as a measure of how well the instrument is tuned or in spec when that data was acquired.

Requirements Checklist

Comms is "the set of components that transport and deliver information from a source to one or more destinations." In this book, we will elaborate on these checklist items. What is your comm network handling?

- Mission Operations Center (MOC) to ground station
- Ground station to satellite
- Satellite to ground station
- Ground station to MOC
- Ground station to data archive
- Flight Dynamics Facility (FDF) orbital ranging data to MOC
- Ranging to FDF
- Ranging to MOC
- MOC telephone system (!)

A good designer looks at a top-down structured approach:

1. Evaluate requirements
2. Identify solutions

3. Create architectureRequirements

Performance Requirements

Your mission is defined and driven by its requirements. A sample list of performance requirements includes:

- Data rate
- Bit-error-rate (BER)
- End-to-end delay (E/E)
- Link availability
- Anti-jam (A/J) capability
- Multi-path needs
- Security: COMSEC and TRANSEC
- Standardization
- Backwards compatibility
- Spacecraft orbit drivers
- Spacecraft mobility
- User-terminal characteristics
- Anti-jamming requirements

Security & Anti-Jamming

- COMSEC: Communications security, encryption
- TRANSEC: Transmission method (e.g., spread-spectrum or frequency hopping)
- Anti-jamming:
  - Blocking friendly accidental interference (e.g., RFI)
  - Blocking intentional electronic warfare
- Multi-access comm theory (FDMA, TDMA, CDMA, etc.)
- Standardization and compatibility and user terminals (obvious)

Spacecraft Orbits

Key mission-driven requirements, including:

- Coverage area (what the spacecraft can see)
- Slant range aka line-of-sight distances visible
- Elevation angle limits
- ...all leading to viewing time

Frequency Allocation

- Limited by hardware
- Regulated by government agencies:
  - FCC (Federal Communications Commission) for industry and amateurs
  - IRAC (Interdepartmental Radio Advisory Committee) for military
  - For other countries, their FCC equivalent
- IARU (International Amateur Radio Union) for amateurs
- ITU (International Telecommunications Union) sets standards
- World Administrative Radio Conference (WARC) advises on standards

Physical Constraints

(aka hardware limits)

- Antenna size
- Transmit power
- Cost
- Channel constraints (air, wire, fiber, space)
- Atmospheric absorption
- Ionospheric attenuation
- Rain attenuation

- Foliage (!)
- Dust attenuation (and volcanoes)
- Solar activity

Architecture Choices

- Single ground station to spacecraft (s/c) (e.g., NASA White Sands)
- Linked ground stations (e.g., NASA's Deep Space Network (DSN) of ground antennas)
- Both: One or multiple ground stations using satellite repeaters
- Crosslink network or constellation of comm providers (e.g., Iridium)
- Multi-access: Multiple ground→Satellite→multiple s/c (e.g., NASA's TDRSS system)

How do you choose from all these checklist items? You take your mission inputs—options, requirements, existing infrastructure, and experience—as your input data. These all lead into your feasibility analysis, the result of which is the set of operational, technical, schedule, and cost limits. You analyze these and generate outputs (aka solutions) that in sum yield your objective architecture.

Comparing a Space-Based Network to a Ground-Based Network

NASA's Tracking and Data Relay Satellite System (*TDRSS*)

- Eight satellites currently in service; three geosync.
- TDRS available for NASA use providing; 24-hour coverage for any satellite.

- Primary ground station is White Sands Complex (WSC) with Network Control Center at Goddard Space Flight Center (GSFC).
- Each TDRS has an S band, Ku band, and Ka band (second and third generation only), forward, return, and telecommunications (source: TDRS homepage (*http://tdrs.gsfc.nasa.gov*)).
- Can support 32 satellites simultaneously!
- Over 300 Mbits/sec capability; 2-48 Mbit/sec avail to ind. missions (S-band only is <6 Mb/s; both S- and Ku-band > 6 Mb/s)

TDRSS cost information is available at the Federal Register and at Instructables (*http://bit.ly/1PgmoRL*), with rates from $10-$131 per minute. This requires a TDRSS-suitable transponder, which tend to be too large and power-hungry for CubeSats. 2015 data indicates TDRSS will phase out per-minute rates in favor of percentage-of-use.

NASA's Deep Space Network (*DSN*)

- Trio of Goldstone, Madrid, and Canberra ground stations, linked to JPL's Deep Space Operations Center (DSOC).
- Each has a 70-meter and 26-meter antenna, and multiple 34-meter antennas.
- Can support two satellites at a time (simultaneously).
- Currently limited to 6 Mb/sec downlink and 25 Mb/sec internal, upgrading to 150 Mb/sec downlink (3-band) and 100 Mb/sec internal. Uplink rate is 2 kb/sec (!) (source: Space Policy Online (*http://www.spacepolicyonline.com*)) (see also ESA's ESTRACK, the Soviet DSN, the Indian DSN, and the Chinese DSN).

Near Earth Network

- NASA ground stations plus some commercial partners
- Maintained by GSFC

- Part of NASA's Space Operations Mission Directorate (SOMD)
- Located at Wallops Island (VA), McMurdo (Antarctica), Merritt Island (FL), White Sands (NM), Svalbard (Norway)

The NEN provides these services to customers with satellites in low Earth orbit (LEO), geosynchronous orbit (GEO), highly elliptical orbit (HEO), and Lunar orbit and to missions with multiple frequency bands. Customers are both national and international, government and commercial entities, NASA (Earth Science, Space Science, and Human Explorations missions) and non-NASA (source: Space Comm and Navigation (*http://spacecomm.nasa.gov*)).

2/Radio

Transceiver Concepts

Communications hardware is the same concept as sensor hardware from Book 3, *DIY Instruments for Amateur Space*. Sensors read in energy—light, electric fields, temperature, vibration, any source of energy—and convert it to an analog voltage, which is then digitized by your computer chip. Comm transmission runs this the other way. It takes a digital data value, converts it to an analog voltage, then pumps it out through an antenna as energy —in this case, as radio waves.

Receiving signals is even easier. A comm receiver is just a sensor. It picks up radio signals and converts them to data. The only difference between a general radio wave sensor and a comm antenna/receiver setup is that the comm setup is tuned to only pick up specific radio wave frequencies, and that the data that comes in has a prespecified format that carries more information than simply *radio wave was here*. There is no additional complexity in terms of hardware or use.

Sensors require (a) a sensor and (b) some connection, be it serial, I2C, or other port, to hook it up to your CPU. Radio comm requires (a) a transceiver and (b) some connection, usually a serial port, to hook it up to your CPU.

A transceiver is a radio device that both transmits and receives, hence the name: trans + receiver = transceiver. We'll cover some standard devices out there. You can get and use separate transmission and receiver devices. In principal there is no difference between using separate chips versus a single chip transceiver. However, given that two chips require twice the wiring, twice the power, and twice the time to integrate, we recommend considering this carefully. We will discuss dual-band use shortly.

One area where people often use just a remote transmitter is for model rocket recovery. Companies sell tiny battery-powered transmitters that send a constant low-power signal so you can find a model rocket that flies into trees or out of sight. In that scenario, you manually start the transmitter just before launch, and manually turn off the transmitter when you have recovered the rocket. Similarly, high-altitude balloons usually use low-power radios that transmit their GPS coordinates to a ground station, without expecting a return. In such cases, a transmitter-only setup is acceptable.

For satellites, both regulations and simple common sense require that you be able to turn off your transmitter if it is overlapping or causing interference to other radio users. Therefore, you need a way to receive signals on your satellite, if only to be able to fulfill the requirement of telling your radio to start and stop transmitting.

As with sensors, there are a variety of different makes and models for transceiver chips. The primary issue for you as a satellite builder is a matter of ensuring you choose the right part, one that is well documented, so you can integrate it into your setup.

Wavelength and Frequency

Radio signals are specified by either their frequency (for example, 433 MHz) or their wavelength (for example, 2 meters). You can calculate any frequency given a wavelength, and vice versa, via their relationship: frequency multiplied by wavelength equals the speed of the signal. In the case of radio, that speed is the speed of light (radio signals are photons, remember), which is 3 x 10^8 m/s, and is commonly abbreviated by the letter *c*. Using simple algebra to move the terms around, we see that *(frequency = wavelength / c* and *(wavelength = frequency / c*.

Wavelengths are usually bundled into bands, with a given comm network or hardware set described by its band range:

- S-band = 2025-2300 MHz
- Ka-band = 18.1-32.3 MHz
- X-band = 7145-8500 MHz

Dual Band and Modulation

A single transceiver both sending and receiving can be full duplex or half duplex. Half duplex means it can send or receive, but not both at the same time, sort of like a walkie-talkie or old NEXTEL push-to-talk phone. Full duplex means it can simultaneously send and receive, like a phone. On the ground, you should always be able to both talk to and hear your satellite, either with a full duplex transceiver or by using two separate radios—one to listen, one to send.

Onboard the satellite, whether to go with a full duplex onboard transceiver, a half duplex transceiver, or two separate transceivers is a decision that has to made on the basis of several factors, including the frequencies you are using, the amount of power available, and the hardware you can build to fly. One common operations design is to have a chosen frequency (such as 440 MHz) and have a single transceiver on the satellite that is set to listen at all times, then transmit only when ordered. An equally common design is a satellite that is more or less constantly transmitting on one band, while always listening for possible commands on another band.

The two bands—sending and receiving—can be the same frequency, or different frequencies. For example, your satellite might be sending data at 437.505 MHz for downlink while listening at 145 MHz for uplink. To complicate matters, a given frequency may be using different modulation methods, such as FM for receiving versus Single Side Band (SSB) for transmitting. Both methods use the same frequency, but modulate the signal differently, and require your hardware be capable of handling that particular choice of modulation scheme. SSB is common because it provides better power efficiency relative to FM for a comparable signal. At a simple level, you need to ensure that your hardware on both ends is matched—if the satellite transmits in SSB, your ground station receives SSB. While you can go deep into modulation theory, at this stage, just consider it (like frequency) a toggle that you decide on based on your allocated frequency and satellite hardware available.

Radio versus Light

Can you communicate with light instead of radio waves? Actually, radio waves are a form a light. Light varies in wavelength from long wavelength (what we call radio), through medium wavelengths (like infrared and visual light), up through the high energy short wavelengths of ultraviolet, X-ray, and gamma rays. You can use any wavelength to communicate.

In this book, we primarily focus on radio communications because it is the easiest and most common implementation for communications. The principles are identical for using lasers or light beams to communicate. You still need a transmitter and a receiver, or a single transceiver to do both. We'll discuss light comm in a separate chapter, just to cover the current state of the art.

Doppler

The number one most important concept that differentiates satellite communications from terrestrial (ordinary) radio communication is the satellite is moving, fast. First and foremost, this means you may have to point your antenna in the right direction, and keep moving it. The second most important effect of this movement is the Doppler shift. From a physics point of view, the Doppler shift is the change in frequency from a transmitter to a receiver based purely on the relative line-of-sight motion between the two. If the two objects are moving towards each other, the frequency shifts higher; if they are moving away from each other, the frequency shifts lower.

Doppler Physics Lesson

The core physics equation, where (for radio) c = speed of light, vr = speed of the receiving, vs = speed of the source, f(orig) is the frequency being sent, and f(new) is the frequency being received, is:

$$f(new) = f(orig) \cdot (c + vr)/(c + vs)$$

You can play around with algebra to yield (when the satellite is at speeds much less than c = 3E8 m/s) that

$f(new) = f(orig) \cdot (1 + dv/c)$, where dv is the relative velocity difference between satellite and station. Given the station is not moving, that means, in essence, dv is the velocity of the satellite. You can rewrite this as $df = f(orig) \cdot dv/c$, which yields the Doppler frequency shift df.

Unfortunately, you have to include geometry—the Doppler equation above is for objects traveling directly towards or away from each other in a straight line. Since an orbit is curved, at any time the satellite's relative velocity to you is a function of the angle between its current direction and you. If the satellite is on an arcing path, you have to take the angle between you and its current direction, Theta, and then the Doppler velocity $dv = v(satellite) \cdot cos(Theta)$. Because the arc is constantly changing its facing relative to you, this means the dv is changing too, and hence df is changing. It is tedious and involves calculus to solve, but software will calculate this for you precisely. Ergo, we recommend you understand the principle and ranges of Doppler, then use existing software appropriately to handle it.

This is why an ambulance siren appears to rise in pitch as it approaches, then drops in pitch as it departs. The Doppler shift affects any wave, including sound, radio, and light. Since we are communicating using radio waves from a rapidly moving satellite (7 to 11 kilometers per second for most orbits), any transmission from your satellite will deliver a Doppler shift as it moves *towards* you (from the horizon towards directly above) then *away* from you (from directly above to the horizon). This will change the broadcast frequency, depending on the mission, anywhere from 1 KHz to 5 Mhz in frequency.

In theory, this means you need to constantly tweak your ground station hardware to be sure you are listening at the exact predicted frequency the satellite signal has. In execution, this means using software to handle the calculation for you. An excellent discussion of Doppler is in *The ARRL Satellite Handbook* from the National Association for Amateur Radio, subtitled "Explore, track and operate ham radio satellites."

You should buy that book as a companion to mine. The focus of their book is not building and operating satellites, but understanding and using existing amateur ham radio communications satellites. Table 2-1 is a sample table of a typical LEO Doppler shift for a satellite transmitter broadcasting at a steady 145.800 MHz, as seen from the ground.

Remember too that your "steady" ground transmitter will experience a Doppler shift from the point of view of the satellite, exactly equal to the Doppler shift in the other direction. Doppler doesn't care which object is moving. Therefore, your station broadcasting at (for example) 145.800 MHz will be heard by the satellite with the same frequency shifts seen below (from the ARRL Satellite Handbook).

Table 2-1. *ISS Doppler, low-elevation pass, transmitting at 145.800 MHz*

Time	Azimuth (degrees)	Elevation (degrees)	Frequency (MHz)
03:57	307	0	145.804
03:58	350	10	145.803
03:59	0	8	145.800
04:00	11	9	147.798
04:01	20	5	145.797
04:02	30	0	145.795

Note the Doppler shift is not linear, but changes more rapidly as the satellite is in the middle of the pass (above you). It is also not the same for each orbit; as the satellite has different peak heights ranging from *above you* one orbit to *it peaked to the left of me* next orbit, so the exact Doppler will be different. Again, when using bands in the 145-435 MHz range (2 m to 70 cm bands), the typical maximum shift will be +/- 3-10 kHz, depending on frequency, and that can frequently be handled by hardware with a sufficiently wide bandpass. At higher frequencies—1 to 10 GHz—the maximum Doppler increases from 30 kHz to over 200 kHz.

The ARRL Satellite Handbook goes on to show other cases, such as high- elevation passes, and discuss them in depth. Your communications design must be able to handle the Doppler shift between your satellite and ground station. Modern transceivers are very flexible in terms of having a broader range of sensitivity, and software-driven communications can be set to the expected shifted frequencies.

One axiom of good radio practice is to never use more broadcast power than you need. With regards to Doppler, for receiving, using a wide bandpass is fine. When transmitting to your satellite, however, you will need to better control your radio usage. You should broadcast as close to the appropriate frequency as possible when commanding, because you are sending radio signals out and too wide a bandpass on your broadcast can disrupt other users. For uplinks and commanding, then, correct for Doppler. Doppler correction (for transmission and receiving) can be handled by many software packages.

Doppler Example

For example, a study of the VO-52 LEO satellite showed the basic 145.86 MHz signal had a typical Doppler of 145.858 MHz to 145.864 MHz during a pass that covered an elevation of 0 to 90 degrees, then drops down to 25 degrees at loss of signal (*http://www.qsl.net/kp4md/doppler.htm*). With a receiver able to cover more than 0.5 MHz of signal, the Doppler is already managed without requiring additional tuning. The ISS has roughly a 3.5 KHz Doppler shift, while most filters allow for 5 KHz of deviation. If you don't match the exact frequency, your power and efficiency are lower (just like a slightly staticky radio channel), but you can still receive signal.

Interference

Although you may worry about radio noise from cell phone towers, WiFi networks, commercial FM radio, fluorescent lights, or other things that put out some radio signals, it is unlikely any of these will interfere with your satellite communications. Commercial devices each are assigned a frequency range, specifically so they don't interfere with other devices. This is why using a Bluetooth device in your car won't ruin your FM radio reception or block the GPS signal to your navigator, for example.

Electronics of any sort do have a tendency to create radio signals because any moving electric field or current can induce a signal in a long antenna-like wire at a distance. This is the reason for the little FCC tags devices have, as proof that the manufacturer built the device well enough, grounded enough, with an appropriate case, so that the device will not be an accidental source of radio signals or interference. Even devices like microwave ovens (microwaves are a form of radio!) are FCC-approved to not spill signal outside of the device itself.

Interference is not just accidental, but can occur when your legitimately functioning device nevertheless conflicts with an existing system. For example, the Tropical Rainfall Measuring Mission satellite has to partially power down its active radar instruments while over Australia, because they overlap the frequency reserved for civil use in that country. CubeSats and other nanosatellites have to be careful, since they often are broadcasting at the same (or similar) frequency as other CubeSats. When that happens, one CubeSat's legitimate signal is interference from the point of a view of a nearby CubeSat. In such cases, depending on priority and negotiation, one satellite may have to power down to allow the other to communicate.

Part of your final hardware build must be designed to not create interference between your transceiver and your payload, and between your payload and any nearby objects, or your transceiver and any nearby objects. Designing is not enough, so you also need to test your final design to ensure it is not creating interference or static while operating. A "noisy" satellite is not just less effective, it is a licensing violation and can result in your mission being ordered to shut down.

ADC

To talk to other humans via radio, you can just use a speaker and a microphone. To talk to satellites, you need to digitize the radio data you're sending. This requires analog to digital conversion (ADC). Radio waves, like sound waves, are analog—they carry data in a mix of waves, each with amplitude and frequency, that overlap to form the signal you hear. Digital data is a series of 0s and 1s that samples the analog signal and makes its best discrete approximation to it. If radio is analog and we are using it to send digital data, we need an ADC (and its counterpart, a DAC—digital to analog converter). With good hardware, this approximation is darn good—think of the high quality digital pictures your camera can take.

Early ADC used a terminal node controller (TNC) that included a modem to convert radio tones into digital data. Some rigs use your computer's sound card inputs to receive the radio, since sound cards already include a *microphone in* jack that you can plug your radio into. In both cases, you have your radio transceiver, which converts radio waves to sound and vice versa. The TNC or sound card (plus software) converts that sound to a digital representation (and vice versa). You can consider both a *black box* that magically handles the transceiver-to-computer interface.

Newer radios now include USB cables that often do the ADC within the transceiver, converting the radio signal directly to digital without having to run it through a speaker as sound. These plug (obviously) into the USB port of your computer, and are just the latest iteration of a black box to handle the ADC step. You can also buy USB radio *dongles*, small radio receivers that plug into your USB port and serve as both radio receiver and ADC together.

For TNCs and sound card usage, and radios with USB cables, you use the system as a two-step process. You still tune to your chosen frequency using the radio itself; the computer serves just to handle the data packets. Packets (which we'll get into later) are bursts of information that your radio sends, that include encoded data (instead of voice). Unlike voice, *listening* to a packet just sounds like random tones, but those tones can be decoded by anyone at the other end who also has a radio with a TNC, sound card, or USB connection.

Most packet formats send ASCII text, kind of like the SMS or tweet of the radio world. Anyone receiving it can decode the packet to reveal both sender and message. You need software to encode or decode packets. Some radios (especially handhelds with digital displays) have the software built in. For your home PC, there are several software packages available depending on the make of your radio and what you need to do.

Sending Code

In amateur ham radio, all packets are sent in the clear without encryption; one rule of amateur radio is you cannot *hide* your signal. Part 97 of the Title 47 rules for hams prohibits, among other details, "messages encoded for the purpose of obscuring their meaning, except as otherwise provided herein." Encryption is hiding or altering the signal itself, while ciphers are using secret cues in the data that only the target knows. Things like WEP on your WiFi router, HTTPS with websites, or SSH to remotely connect to a computer are methods of encryption, as are deliberate encryption encodings like RSA and PGP. Ciphers, in contrast to encryption, are ways of reworking your actual message to be code, such as using a Caesar cipher (A means M, B means N, etc.) or using coded terms ("I want milk" actually means "I want guns," to a recipient in the know). There are exceptions, for example Part 97 includes "Telemetry transmitted by an amateur station on or within 50 km of the Earth's surface is not considered to be codes or ciphers intended to obscure the meaning of communications." There is also some fuzziness (such as the above *milk/guns* example) and ciphering is rarely policed. Remember, however, that amateur radio spectrum is granted under the assumption we share it as a common resource, not a private project. If you require encryption, commercial comm providers can allow that.

Packets

In general, your satellite data feed will start with plain data from either the bus (recording HK data) or the instrument/payload (science data). To this you add a timestamp so you can sort and use your data later. Then you encode it, optionally compress it, apply any encryption and/or authentication, then add in checksums for error checking. Each step adds to the data size, but also makes the data more usable. This creates your data packets.

The difference between raw data and data packets is the same as the difference between *a random binary file of numbers* and *a jpeg image, or a web page, or a PDF file*. The first is just raw data, the second is usable data. Data packet formats exist to provide a universal standard for ensuring that your satellite's collected information is readable from the ground.

Packets are defined as chunks of data containing up to 256 characters. They include the callsign of the sender as well as a callsign or address for the intended recipient, and optionally they include error checking. Error checking depends on the packet format you use, and can be as simple as a checksum or something more involved. The purpose of error checking is to let the recipient tell whether the packet received is whole and complete. An example of a simple checksum for a string of data is the parity bit. You add the parity bit to the end of the data, and it's either 0 or 1. Since digital data is a series of 0s and 1s, the parity bit method counts the number of *1s* in the data and sets itself to either 0 or 1, depending on whether the number of *1s* was even or odd. When you receive a message, you recompute the parity bit for the received message and compare it with the parity bit that was transmitted. If they don't match, if this last bit doesn't match the data, you know there was a transmission problem and at least one bit got flipped. More robust error detection methods exist and are part of the different packet formats.

Historically, ham modes have included the early radioteletype (RTTY) standard and, later, the trio of AMTOR, PACTOR, and G-TOR. These modes are supported by most software. The *TOR modes use automatic repeat request/query (ARQ) for error control, a method that basically has the transmitter retransmit until the receiver finally indicates it received the data without error. PSK31 is an HF digital mode that is often used for the sound card–PC method of connection. Slow-Scan TV (SSTV) is a mode specifically designed to send pictures.

Just as Internet data tends to use the *TCP/IP* format, most satellite data uses either the AX.25 or CCSDS packet format. These packet formats include some degree of timestamp, identifier, and checksums. As openly published standards, this means your data probably will work with existing ground software if you use either standard. Standards improve efficiency and interoperability. In DIY terms, this means you should always use an established data packet standard unless you happen to be an expert in the field. If you're reading this book, that probably translates to *use an established data packet standard*.

Data Modes

A canonical list of data modes, taken from Wikipedia (*http://bit.ly/1PgtfuF*), notes "most amateur digital modes are transmitted by inserting audio into the microphone input of a radio and using an analog scheme, such as amplitude modulation (AM), frequency modulation (FM), or single-sideband modulation (SSB)." Their list is:

- Amateur teleprinting over radio (AMTOR)
- D-STAR (Digital Data), a high speed (128 kbit/s), data-only mode
- Hellschreiber, also referred to as either Feld-Hell, or Hell
- Discrete multi-tone modulation modes such as Multi Tone 63 (MT63)
- Multiple frequency-shift keying (MFSK) modes such as FSK441, JT6M, JT65, and Olivia MFSK
- Packet radio (AX.25) and the Automatic Packet Reporting System (APRS)
- PACTOR
- Phase-shift keying: PSK31, QPSK31, PSK63, QPSK63
- Radioteletype (RTTY)

Any packet format is just like an envelope to send a letter. It doesn't matter what the contents of the letter are, only that it (a) fit into the envelope and (b) has an address and return address so recipients know where it came from and who should get it. The current default digital packet format, especially for satellite and high-altitude balloon (HAB) use, is AX.25. Being a standard, this is also what most software supports. Each AX.25 message includes the call sign of the sender (or satellite) plus the call sign of the destination station, so each message can be uniquely identified and, if needed, relayed to the appropriate user.

In addition to AX.25, there is CCSDS, a new standard that is being promoted especially for satellite use. CCSDS is more accurately the CCSDS Space Packet Protocol, as CCSDS stands for the Consultative Committee for Space Data Standards. The *blue book* at ccsds.org defines the exact packet format. CCSDS packets are specifically defined for space use, allowing for the space equivalent of FTP connections (called FDP), for retransmission of data, and for effective data transmission along the often noisy, lossy radio channels.

When would you use CCSDS instead of AX.25? If your software supports it. For example, CCSDS has tools such as the CCSDS File Delivery Protocol (CFDP), which lets you treat your satellite just like a remote directory from which you you can uplink or downlink data.

Packet data transmits at rates of 1200 bps to 9600 bps, so it is not very fast, but it is efficient and it is the fastest connection that can be supported by the radio frequencies and hardware used. Put another way, you could optimize a new packet format that yielded higher bandwidth than AX.25, but then you would no longer be compatible with any existing software or with other users.

As you may gather, AX.25 and, optionally, CCSDS are the primary items you should focus on. The terminology can be tricky; for example, AX.25 is the protocol for the packet mode of communication, which can be used on a variety of HF and UHF frequencies, and is independent of how your hardware actually modulates the signal. So you'll read an item like (from Wikipedia's SkyCube entry) "the SkyCube radio emits periodic beaconing pings which contain 120-byte messages from the Kickstarter backers. These pings are transmitted at 915 MHz, using the AX.25 protocol at 9600 baud with BPSK modulation, with a callsign of WG9XMF."

By now, hopefully you can parse that out as the satellite transmitter hardware is using the frequency of 915 MHz, and likewise the radio hardware is using BPSK modulation in sending the signal. Once your ground station receives this, the data itself uses the AX.25 packet format, and the data rate is 9600 baud, loosely 9600 bps (though baud != bit, but let's not get into that here). Having translated the data into human-readable format, each data item contains the callsign WG9XMF so you can identify that they are from SkyCube, and the actual information inside each message is likely 120 ASCII characters (assuming they are using ASCII, where 1 letter = 1 byte). That's our three layers—hardware (frequency and modulation and data rate), data (data rate and packets and protocols), and user (informational content sent or received).

More to the point, you shouldn't have to deeply understand either AX.25 or CCSDS in order to use them. They are akin to the GIF versus JPEG formats when sending images—the user doesn't need to know the detailed workings, only the software needs to support it. Your software should understand the format, and you need to ensure your satellite's processor is formatting the data properly. In short, the data packet spec matters to the code, but does not fundamentally change your operations or your design. Much as you would choose either *TCP* or *UDP* when making an application to connect two devices over the Internet, then use a software library to handle the actual conversion, you should pick AX 25 or CCSDS, then use the appropriate software and libraries to manage your data for you.

Core Flight System (CFS)

The Core Flight System (CFS) (*http://1.usa.gov/1PgtPZo*), an open source NASA-coordinated command-and-control system that includes both ground and satellite software, may end up being a game changer. CFS uses CCSDS. You install CFS on your satellite CPU, then you install CFS on your desktop PC, and you have a working satellite OS and ground system. Currently CFS has only a dozen flight software modules, to manage commanding and data handling. The open source public release was March 2015, but CFS is already used on existing missions such as the Global Precipitation Measurement (GPM) mission and the Lunar Atmosphere and Dust Environment Explorer (LADEE) mission.

APRS

The Automatic Packet Reporting System (APRS) is one of the most brilliant aspects of ham radio. APRS is a worldwide networking of radio users, automated stations, repeaters, relays, and the Internet that serves to capture and relay APRS-identified packets to anyone, in near real time.

If your satellite (or high-altitude balloon) is transmitting an APRS packet, you in essence now have a worldwide ground station that you can access via Internet, without even owning a radio yourself. APRS uses 144.39 MHz (in North America), 144.80 MHz (Europe), 145.175 MHz (Australia), and 145.825 (Space). APRS is intended to provide local information including status, weather, location, and other relevant information. Details are at *http://www.aprs.org*, and you can see archived messages in the APRS-IS (APRS Internet System) database at *http://www.findu.com*.

As written by APRS founder Bob Bruninga, WB4APR, *"It is a two-way tactical real-time digital communications system between all assets in a network sharing information about everything going on in the local area. On ham radio, this means if something is happening now, or there is information that could be valuable to you, then it should show up on your APRS radio in your mobile."* APRS uses AX.25 packets with specific protocols and formats, which differ depending on whether you are sending (for example) weather data, GPS coordinates and a position report, or other data.

Previous CubeSats such as PCSAT-1 and 2 and BricSat have used or intend to use APRS to report their location, status, and basic data such as temperature or electric/magnetic field measurements. Further details are at USNA CUBESAT Notes (*http://www.aprs.org/cubesat.html*) and the Cubesat Comms Design Page (*http://www.aprs.org/cubesat-comms.html*).

One fun aspect of APRS is that, even without a radio, you can access the data via the APRS-IS databases. In addition to the above resources, visit the APRS Satellite Tracking and Reporting System (ASTARS) (*http://www.aprs.org/astars.html*) to track and predict existing satellites as well as view their APRS messages.

Software-Defined Radio

Software-defined radio (SDR) is perhaps the most relevant hardware innovation for receiving satellite information. The long and short of SDR is that you can buy an off-the-shelf $20 USB dongle, and get performance equal to $500+ hardware radios of yesteryear.

An SDR dongle is a radio receive that covers a wide band—typically 24 MHz through 1.7 GHz, which includes most satellite bands. The unit plugs into the USB port of your computer, and you must provide a driver and software to use it. You also plug an antenna into it—the better the antenna, the better your reception.

Anyone can use an SDR and you do not need any license to receive any signal a satellite transmits. SDRs are a cheap way to make any laptop or USB-equipped tablet serve as your *satellite scanner* to listen for data. While this isn't a base station if you wish to command a satellite—that requires transmission capability, and a Ham or other license—SDRs are awesome for getting into listening to satellites.

To buy one, search on your favorite search engine or on eBay for *SDR USB RTL* and you can find viable units under $25US. The unit includes both an ADC (analog-to-digital converter) to translate from the antenna RF signal to computer digital units, and a tuner to allow you to sub-select which frequencies to look at. A typical unit might use an RTL2832 ADC chip for the former, and an R820T or similar tuner. The primary issue when buying an SDR is to check whether your computer supports it. The general spec is the *RTL282* chip, which most Windows, Linux, and Mac computers can support.

The tuner itself usually covers a range of 3.2 MHz, which means that, for example, you can pick up a 145.00 MHz signal as long as it is broadcast in the range of 143.4 to 146.6 MHz. This is useful because satellites are moving, and in the Doppler section we noted that motion induces frequency shifts. A decent span means that, even if the frequency is shifted, you can use this to capture it without having to manually adjust your frequency.

The trade-off of a large bandpass such as 3.2 MHz with a core central frequency of 145 MHz (for example) means that you risk catching other stations in your listening. By analogy, a typical bandpass for a home FM radio is about of 0.5 MHz. Radio stations in metropolitan areas should keep at least 0.5 MHz apart from each other: a station like *98 Rock* in Baltimore, which transmits at 97.9 MHz, should not have any neighboring station between 97.4 and 98.5 MHz. Below 97.4 or above 98.4, a nearby station will not interfere with 98 Rock.

This could be a problem with an SDR's larger bandpass. Tuning into 97.9 with a 3.2 MHZ bandpass means you'll not only be receiving 97.9, but every station between 96.6 and 99.5.

Fortunately, satellites using similar frequences are rarely near each other. In that case, a wide bandpass means that you can catch your one satellite even if it experiences Doppler shifts as it broadcasts, because ideally it is the only item in the sky in the direction you are pointing. If multiple satellites exist in the same zone of the sky as you are pointing, all using similar (but slightly different) frequencies that differ only due to Doppler broadening, you likely would not be able to distinguish them even without Doppler. Satellites are unique items, and we rely on this in listening. So, an SDR with a wide bandpass is an ideal tool for listening.

Flexible hardware implementations for stand-alone radio gear that you wish to integrate with your computer also exist. AMSAT notes a computer sound card can handle +/- 24 KHz sampling around a frequency (a 48 KHz bandpass) that typical hardware provides (QST magazine, April 2014, ARRL, pg 33, 'A Tiny Python Panadapter"). The key trade-off is that a stand-alone radio receiver will have higher signal and lower noise, but with typically a smaller bandpass, compared with the tiny USB SDRs mentioned.

SDR only listens; to date none of them transmit signals. You can use an SDR to receive, but still need a ham set (even a handheld Baofeng) plus the appropriate licensing, to transmit to your satellite. SDR is also available as an Android app for use with tablets, smartphones, and other devices, using software such as *SDR Touch* or *SDRAnywhere*. SDR is constantly dropping in price and improving in performance and in software availability, and is the easiest way to get started (at the very least) listening to satellites from your laptop.

The International Space Station for Hams

The International Space Station (ISS) actively supports and participates in ham radio. Anyone with a radio can listen, and anyone with a license can transmit to the ISS in the appropriately licensed bands. NASA also supports voice communication with the ISS via amateur radio, especially for teachers and classrooms (search for ARISS on NASA's website (*http://www.nasa.gov*)). Details for ARISS aka Amateur Radio are listed on the ISS (*http://www.hamuniverse.com/space.html*); here are the ARISS (ISS) frequencies:

"These frequencies are currently used for ARISS general QSO:

- Voice and Packet Downlink: 145.80 (Worldwide)
- Voice Uplink: 144.49 for Regions 2 and 3 (The Americas / Pacific)
- Voice Uplink: 145.20 for Region 1 (Europe, Central Asia and Africa)
- Packet Uplink: 145.99 (Worldwide)
- Crossband FM repeater downlink: 145.80 MHz (Worldwide)
- Crossband FM repeater uplink: 437.80 MHz (Worldwide)"

More details are available from *ariss.org* and other sites.

3/Full Ground Station

What is your mission producing? Why, it's producing Mission Data. This includes health and safety telemetry—which your customers couldn't care less about. Primarily, they want the payload or instrument data. Processing this data and generating useful technical or scientific data is what you (or your customer wants).

Missions often split into a Mission Operations Center (MOC) that operates and maintains the health of the satellite and handles communications uplinks and downlinks, and a separate Science Operations Center (SOC) that receives the payload/instrument data from the MOC and distributes it to customers. Think of the MOC as the engineering side and the SOC as the science side. As an Internet analogy, think of the MOC as the server and the SOC as the web pages being served.

Whatever instruments or payload you are flying, you want your data like you want your TV—more of everything.

- High resolution
- High cadence
- Multiple wavelengths/multi-spectral

Expecting it to be like a Discovery Channel movie, users expect raw, unedited high-resolution images at 32 fps using multiple cameras. What they get (they feel) is tiny VGA snapshots from a cheap digital phone.

Downlink Only

At this point, we're talking about downlink only. The MOC has no control over the data directly; however, actions by the MOC can affect the:

- Quality (fidelity) of the measurements
- Completeness of the data

Data levels are defined as:

- MOC has level 0.
- SOC may take level 0 and produce either Quicklooks or level 1A.

There is an eternal conflict between transmitting all raw data to Earth, or having the satellite select only the best data to transmit down. Onboard processing of the data reduces the telemetry needs and ensures more data is received; sending raw data ensures that the data sent has more information and encourages unexpected data discovery.

At a government or corporate level, data handling often involves one or more DAACs (Data Active Archive Centers), or an entity like NSSDC (National Space Science Data Center). For your mission, your MOC, SOC, and DAC may all be just *the laptop running your mission*, but the roles still remain distinct.

Science data does not just include "the bits sent by the instruments," but also:

- The raw data from the spacecraft
- Attitude and orbit information about when and how the data was taken
- Processed data (Level 1, 2, etc.)
- Data quality flags (from Ops)
- Calibration level and version info
- Other metadata (e.g., planning info)
- Ownership info

Operations Choices

Your "Concept of Operations Concept" or ConOps is how you plan to operate your satellite, and must be defined before you start building anything. So let's build a ground system (on paper first)! Here are the steps to design a system:

1. Identify mission needs
2. Determine ops tasks
3. Identify how to do it (and whether we already can do it)
4. Do key trades analysis
5. Develop scenarios
6. Make timelines
7. Assess resources needed
8. Map data flow
9. Identify team
10. Assess costs
11. Derive requirements
12. Make tech plan
13. Iterate

Notice "pick a set of choices randomly and just go with it" is not a valid option. But to do this, we need to look at the core assets involved in a ground system.

Start with the Mission Payload

We assert three categories for defining a mission type and objectives:

1. Trajectory aka orbit
2. Type of payload: Comm, Science, Navigation, Remote Sensing, R&D, Exploration, Tech Demo, etc.
3. Payload complexity

Now we assume that has been defined, so we can work with the tangible assets we are flying:

- Pointing detector(s) and/or telescope (e.g., HST)
- In-situ detectors (e.g., radiation monitor)

- Built-in lab (e.g., bio)
- Space-based factory (e.g., crystals)
- Comm system/repeater (e.g., OSCAR)
- Navigation system (e.g., GPS)
- Weapon(s) (e.g., space laser, EMP)
- Tech demo (e.g., ion drive, solar sail)
- Astronaut(s) (or scientists or space tourists)
- Deployables and probes (e.g., sails, landers, etc.)

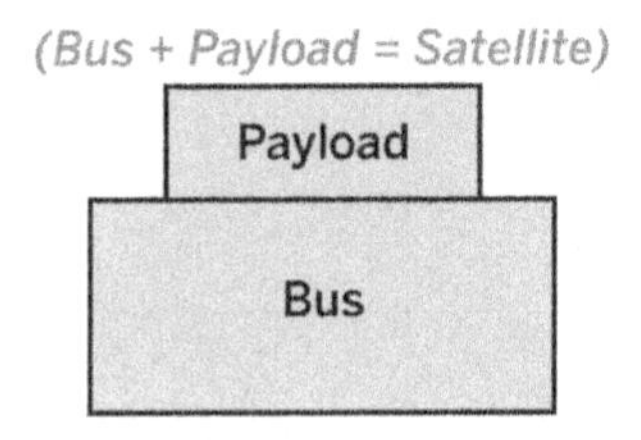

We need a bus that supports the payload. The bus + the payload = the satellite. The fundamental components of a bus are:

- Power: batteries, solar cells, perhaps an RTG
- Computing: CPU and data storage
- Communications: Transceiver (both send and receive) and antenna
- Attitude and maneuver: Gyros, thrusters, magnetic torque, etc., also star trackers, etc.

Note that "Communications" is part of the bus, but that our entire *command and control* for operations has to include the entire bus. We will monitor the bus—physical temperature, status of power, status of any systems, status of instruments—as well as download data from the instruments and send up commands. So our operations concept has to look at the entire satellite.

We need to know what the orbit is, as that partially defines your comms by showing what Earth-based locations and what relay satellites potentially can see and communicate with your satellite. Once you tie the orbit to the rotating Earth, you can then begin to map out when and how often you can directly communication with your satellite:

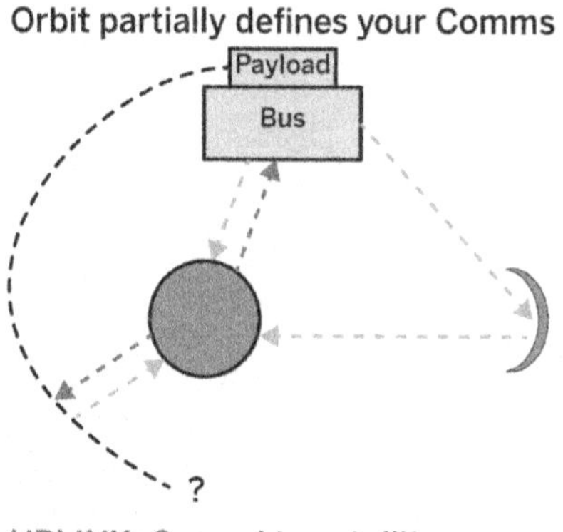

UPLINK: Ground to satellite
Sending commands or data to a satellite

DOWNLINK: Satellite to ground
Receiving data from a satellite

Typical orbits include LEO (circular or eccentric), MEO (circular or eccentric), HEO (circular or eccentric), GEO, halo, helio, or planetary/trajectory. The orbit defines the potential comms.

Comm Quantities

UPLINK is ground to satellite, sending commands or data to a satellite. DOWNLINK is satellite to ground, receiving data from a satellite. The dueling communications quantities are:

- Number of times you can contact it per day
- Duration of contacts
- Amount of data you can uplink per contact
- Amount of data you can downlink per contact

VERSUS

- How frequently it must be commanded
- How much data (if any) per day it generates
- How much data it can store onboard before requiring a transmit to ground

- (If CommSat, uplink/downlink are also required)

"Comms" is not necessarily a single entity, or even unique. Different entities can handle different aspects of the data. One common NASA scenario is that all telemetry goes through a dedicated comm provider like DSN or TDRSS/White Sands. This comm provider routes data to the appropriate entities. The Mission Operations Center (MOC) prepares uplinks and receives all housekeeping (HK) telemetry to assess the health and safety (H&S) of the satellite. Meanwhile, the Science Operations Center (SOC) feeds the MOC with plans for upcoming operations, and also receives Quicklook (thumbnail) science data plus a subset of MOC's HK—specifically, the instrument HK data for monitoring instrument H&S. The third entity, the Data Archive Center (DAC), receives all the stored science data.

These three components—MOC, SOC, and DAC—are the core of your ground system. For a small mission, all three entities may be one person in their basement. For a large-scale science satellite, each entity may be in a different country. For a communications provider, there may be multiple MOCs and, in addition, each subscribed user may be the equivalent of a SOC. One common amateur radio scenario is that each user directly connects to relay satellite. User A sends a message to the relay satellite. User B receives a message from the relay satellite. A different *DirectTV* scenario is that the comm network MOC sends commands and data up to satellite, while individual users can (passively) receive data from satellite.

In all cases, we are not just separating by whether an entity has uplink or downlink capability and whether they have access to command the satellite, but also whether they are receiving satellite HK data, payload data, or both.

Ground Network

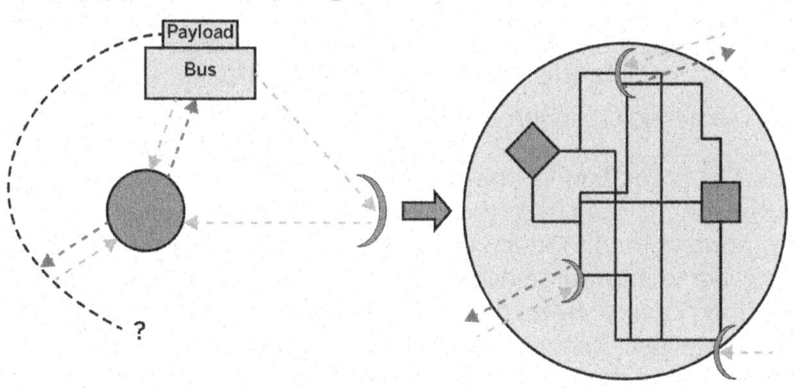

What are the physical ground assets handling the communications?

- Antenna(e)
- Comm ops
- Data distribution
- Data archiving
- End users

The questions you need to answer will help define what your comm capabilities are. Where are they located? Are they fixed or mobile? What is their capacity? When do they operate? Who owns them?

Ranging and Orbit Determination

In addition to items you can control—your orbit choice, the ground stations that will support it—there is an additional component that is often hard for the amateur mission to achieve on their own. The question is how to know where your satellite is. Finding the satellite's position and time is known as ranging, and from this data you can do an orbit determination so as to generate a set of orbital elements that enable you—finally, at your station—to predict where your satellite will be in the near future. Three methods exist to figure this out:

1. You predict based on last known position and velocity (inertial).
2. Your satellite tells you (GPS or equivalent).
3. Someone on the ground tells you (ranging).

Inertial prediction is the least reliable, as deviations due to atmospheric drag, that the Earth is not a pure sphere, and that computers tend to round numbers all lead to the prediction getting worse and worse over time. Having the satellite tell you its position is very useful now that GPS technology is available for LEO satellites. Ranging by US Space Command (formerly called NORAD, and the NORAD facility is still referenced for ranging) is done automatically for every satellite, and if you can match your satellite to the ID which NORAD assigned, this provides a way to get the position on a regular basis.

Sometimes orbit determination is (often?) co-located with your commercial comm provider. Sometimes a third entity such as NORAD/US Space Command handles it. In that case, there will often be a delay before you get the latest information. Sometimes it is up to you, and you must do your own ranging. We'll cover this in our chapter on orbits. It is possible but not recommended; wait for NORAD to let you know where your satellite is.

Ground System Design Products

The end result of all the above questions and your analysis and answers should feed into an overall map and layout of your ground assets, in a tactical sense. The components you will create include:

- Geographic maps (where are your assets?)
- Coverage maps (who sees what when?)
- Timelines and schedules (when do we have to do what?)
- Network maps (how are the ground assets connected?)
- Org charts (what entities are involved, who bosses who?)

- Resource budgets (spreadsheets to balance your comm capacity against your uplink/downlink data needs.)
- Cost budgets (you'll pay for this!)

Now let's look at some of the big picture risks and how to make assessments.

Mission Phases

Your communications needs will change as your mission evolves. Breaking your timeline into phases will help you decide how to ramp up your mission, deliver on your core goals, and potentially extend your mission life or even repurpose your satellite for alternative uses later on.

The pre-launch phase is where you are building and testing your ground system. During this time, only your core team needs to be active, because there isn't actually an orbiting asset to manage. Also, during this period, you may be altering your software and hardware choices as you converge on a good design. As you start to get close to launch, you can begin to deploy your specifications and final software to any partners. This is also the time that all licensing and (if money is changing hands) contracts are resolved. Plan in advance how you will get extra capability if you run into trouble after launch.

In-orbit checkout (IOC) is that period immediately after launch, and extending anywhere from days to weeks, during which your satellite first is released, then powers up, and then begins doing simple operations. This is the single most crucial period of the mission, during which you need as much communication capability as you can get your hands on. For IOC, you can draft volunteers to provide additional ground stations for the short term. The core mission of IOC is to ignore the payload, and just determine if the satellite bus—the power, comms, CPU, and any components that are required to operate to keep your satellite alive—is working properly. During IOC, you want to minimize risk, because any loss is a big cost loss. In this phase, *Failure is not an option*: you have no idea what might fail on your satellite, so you'll want to have redundant coverage.

Next comes the performance verification (PV) phase, in which you now verify your payload. This can take days or weeks. During this time, you may need to tweak payload settings, fix command scripts, or do other ground operations to get the darn thing to work properly. Since IOC confirmed that you can listen to and command your bus, you likely will not require redundant ground capacity anymore. Instead, you can use whatever your core long-term ground plan is to leisurely tweak your payload until you've got it tuned and humming. You still want to minimize risk, so the goal here is to establish *nominal operations*, where nominal means the normal expected level. This is not the time to change things, just the time to get things working as planned.

Next comes the primary mission phase(s), where your satellite does useful things. The primary mission may be as simple as *send a Sputnik ping* or it may have distinct phases such as *(1) go to Pluto then (2) measure Pluto*. This period is the best part of operations, because you've verified that you have a working satellite (from IOC) that is returning valid data (from PV), so now you can carry out your mission. You still want to minimize risks and changes so your mission can go as close to how you planned it as possible. How long this phase lasts is up to you, and can be extended until the satellite's final fiery end.

Optional extended mission phase(s) are when you add or change the mission goals due to your satellite performing longer and better than the primary mission required. This is *extra play time*, and now you can try all the risky ideas you thought up during the primary mission but were wise enough to not try until now. For example, if your primary mission is to collect three months of data and, three months later, your satellite is still alive, you can go into an extended mission where you can keep doing the same thing, or try some new things that might or might not work. The key value of extended missions is that you can assume increasing risk in hopes of higher reward. Your satellite may also have reduced capacity at this point. Instrument or platform capability may be reduced or degraded due to cost (comm, staffing), failures in the platform (especially gyros or power), instrument problems (especially on multi-instrument missions), cumulative degradation and reduced capacity of the platform or instruments, loss of consumables (fuel, cryogen), and change in location (orbit decays, flybys, loss of some ground stations). An extended mission is the art of adapting to the new situation in a way that lets the satellite still do useful work.

Extended missions can include repurposing, where you completely reinvent the satellite. The WISE infrared astronomical space telescope, for example, reached its end of mission when it ran out of its finite amount of coolant for the main instrument; a decade later, it was powered back up as NEOWISE, a near-Earth asteroid spotter, using the secondary capabilities remaining. The solar physics International Sun/Earth Explorer 3 (ISEE-3) mission was later repurposed—and renamed—as the International Cometary Explorer (ICE) comet rendezvous mission once operators realized it had enough fuel to be the first spacecraft to visit a passing comet. If nothing else, perhaps your satellite can be Repurposed as a target for someone's space laser? Perhaps not, in which case you will eventually get to...

The end of lifetime for the satellite, at which point you have to decide how to deorbit it. Deorbiting can be as simple as relying on natural drag forces to force a re-entry, or can involve actively commanding a deorbit mechanism to quickly force the satellite to reenter. Upon reentry, the satellite will burn up in the atmosphere. Mission over! Time to start a new project.

4/Licensing

Overview

Satellite communications technology is very straightforward. Hardware is cheap and ubiquitous, orbital mechanics leads to predictable satellite paths, and software exists for operating your satellite. The primary operational limit that defines this is your orbit, but the biggest hurdle is licensing. Any broadcaster has to have permission from their parent country to use long-distance radio, and the terms of use often restrict the power level, frequency of communication, and allowable uses of that radio.

To bracket the problem, let's first look at the amount of time you have to communicate. If you go for an inexpensive low Earth orbit (LEO) launch, you're orbiting the Earth every 90 minutes and only one-third to one-half of your orbits each day will be over any given location. Let's say you've set up your ground station in Makerville, in the country of Makerland. You *see* your satellite four to five times per day, but each of these *contact passes* lasts no more than 10 minutes. This yields at best three-quarters of an hour for any one ground station to communicate with your satellite. You can expand your operations base. Add more ground stations, and you get more ground time. Include an Iridium-like constellation of relay satellites so at any given time, at least one satellite can see yours, and you have a relay system with high time availability. Go for a higher geosynchronous orbit (GEO) and your one ground station can always stare at your satellite, but that's a more expensive orbit to launch.

All of these are technical aspects of your mission, but they all bottleneck in the stage of *space to ground*. To broadcast anything up or down, you need to have a radio license. Spectrum licensing isn't cheap, and you can't use the amateur bands for a commercial service. Key questions in determining which license terms you are eligible for include:

1. What is your budget (what is *affordable*)?
2. Who is your audience or customer base? Where are they located? What gear will they have?
3. What services are you providing (email versus Wikipedia access versus streaming video, for example)?
4. What sort of uptime/downtime is acceptable?
5. Is this for profit, registered nonprofit, or independent/unlicensed/altruistic?
6. Do you have any contacts/connection either in the country, the region you wish to serve, or similar in case legal or licensing issues come up? . And perhaps most important, what is your goal or intent?

Extended Case: Private Comm Sat

Let's take a sample mission often raised, "Let's build a CubeSat as a private communications satellite." That merely defines the hardware. To really understand how to communicate with it, you need to examine what the goal of the comm satellite is. Is your mission to make the world a better place, to bootstrap local regions so they improve economically, to make a pile o' cash for yourself, to prove the concept is feasible so someone else can invest more money to scale it up, to break censoring by the existing government, to encourage local entrepreneurs, to build an amateur radio community in a region that doesn't have one, to inspire DIYers, to create better educational opportunities, to implement a mind control algorithm to conquer a continent? Each has different drivers and further questions. Look at any problem to be solved by tech or science (such as *cure cancer*) and you have everything from NIH research centers to pink ribbon campaigns to alternative medicine to hospices. What's your stance and what's your angle?

If you are trying to do your mission as a competitive or commercial venture, you run into regulatory issues. If you're trying to do altruistic *power to the people* and give free email access to regions in a nonprofit way, something like Winlink (*http://www.winlink.org*) (email via amateur ham radio) might be the way. In that case, you're making use of existing assets (ham radio and that there are already AMSAT satellite repeaters) and you simply need to get gear out to users, maybe as a sort of Internet café approach.

For giving access to poor regions, flying a satellite might be overkill. Iridium went bankrupt because cell providers realized that, instead of launching comm sats, they could just put up lots of cheap ground stations. I think a ground network of *cantennas* is a better approach. A cantenna is a cheap antenna for off-the-shelf WiFi routers that can reach about 10 miles between nodes (the record is an insane 382 km (*http://oreil.ly/1PgXZeQ*)), but pragmatically, having a bunch of cheap solar-powered ground repeaters would be feasible. Think of them like comm satellites that don't fly. If you want to get them off the ground, put them in a blimp, to increase line-of-sight.

This divergence into "Should I fly a comm sat?" is intended to show that the new space frontier is not the only way to solve global problems. Given that comms and licensing is a major hurdle for satellites, it is worth looking into other approaches for handling comms.

Future Ideas

Is the current model, where each amateur satellite builder also has to make his or her own ground station, the best approach? If I had to predict, I'd say that within four years (i.e., by 2019), someone will build a constellation of satellites specifically for picosatellite and amateur usage. It will likely use radio in space for the satellites to talk with each other, with a shared radio or laser comm downlink between the main space-based relay, and several dedicated ground stations. This is the *cable TV* model—one big *pipe* for the ground-to-space-relay segment, then lots of cheap lower bandwidth radio usage for individual satellites in the space network.

One advantage of this model is you have different regulation for ground-to-space than space-to-space. It isolates the ground operations portion from the individual satellite operations, because each satellite is routing through a single ground station. This means satellite builders don't necessarily need their own ground station, but just tap into their satellite through regular Internet. It is also highly extensible. As amateur satellites go past LEO into lunar orbit and beyond, the network can expand to support this.

A colleague thinks that the existing satellite comm providers will find it economically viable to open their current ground-(via-space)-to-ground satellite phone systems to allow ground-to-space. In this model, you simply subscribe to their existing satellite phone package, and bundle a phone in your satellite. This would be a very elegant solution and, again, obviates the need for your own dedicated ground station.

Already, high-altitude balloon (HAB) experimenters are trying different approaches to comm, beyond ham radio and amateur transmitters. Notable efforts include:

- Commercial off-the-shelf (COTS) Android phones in space, as general platforms but not as comm devices:
  - How NASA got an Android handset ready to go into space (*http://bit.ly/1PgZJEY*)
  - NASA's PhoneSat 2.5 Dials Up Innovation for Smartphone Satellites (*http://bit.ly/1PgZQAy*)
- Using a SPOT GPS position device from a HAB (*http://bit.ly/1R4aNCJ*)
- Iridium GPS-only for satellites (*http://bit.ly/1Ph17HF*)
- Sending text messages from a cell phone on an HAB:
  - High Altitude Balloon Tutorial Cell Phone Transmitter (*http://bit.ly/1PhOTAu*)
  - Tracker tutorial (*http://bit.ly/1R4aPKR*)
- Why cell phones aren't licensed for HAB (*http://bit.ly/1R4aQhZ*)
- Why cell phones don't work from the ISS (*http://bit.ly/1PhO22B*)
- Iridium for HAB:
  - Iridium: Global OTH data communications for high altitude scientific ballooning (*http://bit.ly/1PhO6zw*)

    - ERROR WHEN I CLICK THIS LINK (*https://www.rock7mobile.com/products-iridium-sbd*)
- Iridium on a high altitude balloon updating Facebook and Twitter:
    - Introducing AllAloft – The (to be) Complete Iridium-based HAB System (*http://bit.ly/1Ph0g9X*)
    - Project Hermes: New horizons in satellite communications (*http://bit.ly/1Ph0kGO*)

Space isn't cheap yet—but cheap network access, that's solvable.

Regulatory Solutions

The hardest aspect of satellite communications is regulatory. Radio frequencies are a shared resource. I will present several case studies of how existing picosatellites solved the communications problem. There are three main problems involved with communicating with an amateur satellite. They are, in rough priority of difficulty:

1. Regulatory: do you have permission to broadcast from the satellite, or upload commands?
2. Technical: how do you build a system to communicate?
3. Bandwidth: how often, and how much data, can you send up or get down?

IARU

The International Amateur Radio Union (IARU) is a boon to mankind, and a gatekeeper. They are a boon, because they negotiate and maintain amateur access for radio frequencies suitable for satellite use. The gatekeeper role is that amateurs have to apply for permission to used this shared spectrum.

The website (*http://www.amsat.org.uk/iaru*) discussed the IARU Satellite Frequency Coordination Panel. Their form for requesting coordination does not require deep technical understanding—primarily the proposed transmitting and receiving frequency plan you wish to use, the launch date and duration of your mission, and some basic satellite specifications. You will include your output power, your link budget, and your antenna gain and pattern. You also need to indicate how you will turn off your satellite immediately if requested (for example, if there is interference, or if another higher priority task requires that frequency).

The March 2012 version of their "Amateur Satellite Frequency Coordination Request" form (available at the above web address for free) notes, among other key points, that "Amateurs are expected to coordinate their use of frequencies," "Amateur radio satellites present a special problem because satellites have global effect," and that "Frequency coordination for amateur radio satellites is provided by the IARU."

If the purpose of your satellite is not specifically aligned with *amateur radio*, you may not necessarily be able to use the IARU-coordinated frequencies. For example, if you just want amateur radio access to send down data to your own lab, this is not a benefit to amateur radio in general. However, if your mission includes a component that is relevant to amateur radio users in some component, you may qualify for IARU coordination. The Eagle-2 satellite, for example, had a contest where amateur operators could try to discover the *secret* commanding modes to change the satellite's transmissions, thus qualifying it as a radio experiment (in addition to its primary mission as an engineering test build).

Smartphones

"Why not just use an iPhone?" or "Why not just use satellite Internet?" are common questions. The answer is not technical. The answer is that you are violating the usual license terms, and therefore you have to get permission from your Internet/comm provider.

Most comm satellites are in GEO, which means LEO is not a technical problem. Even for networks like Iridium, which uses a constellation of LEO satellites, comm with another satellite is likely not a technological issue. Instead, it's an issue of terms of service and of cost.

We live in an age where they've put a cell phone tower so Mt. Everest climbers can tweet from the peak. There are no barriers due to technology, but you have to find a provider whose terms of service include *space to ground*.

Early high-alt iPhone customers violated their cell phone TOS to send a signal. In that case, cell phone companies had a legitimate concern—at high altitude, you can be reaching multiple towers and using disproportionate network usage.

High-Altitude Ballooning and Guerrilla Radio

Technically, if you are licensed and at low altitude, you can do short-term radio work without any additional FCC clearance. I know this empirically, through tagging along with a university high-altitude balloon payload.

The gear is dirt simple—a handheld walkie-talkie hooked up to a display to show the packets. The balloon payload was just sending down GPS coordinates, and this was an ideal way to track. Being radio, the repeater network was also in effect—essentially, using off-site radio people to help if we lost our mostly line-of-sight and must-be-close.

Demystifying the Basic Technology

The basic tech to do direct ground-to-satellite and satellite-to-ground radio comm is very simple.

Receiving data

A satellite has a radio transmitter (radiometric, etc.) and a processor (BasicX, Arduino, Atom, see previous books). When commanded to, a satellite sends data to a transmitter with proper formatting. The transmitter converts it to a radio wave and any ground station can pick up the signal.

The basic hardware for most transmitters is as follows. The processor sends a serial signal to a transmitter chip. Literally, *send data* just like for a sensor. The chip is smart enough to take that and convert it to a radio wave. Done! Except you need to attach your ham tag, for ID.

The specifics of satellite comm are best set out in *The ARRL Satelite Handbook*. It includes details from AMSAT, who have been doing amateur satellite communications using satellites they built by hand—for 40 years! So, yes, join AMSAT.

Regulatory and Permissions

The amateur spectrum requires Open Data Standards. Any satellite data using amateur radio has to be sent unencrypted.

Ham Culture

Be aware that most ham radio advocates enjoy the art and use of amateur radio to talk to other hams. I get a bit of heat in that I attend radio events, but have no interest in their main events—ICQs (recording who you talked to and how far away they are), etc. Fox hunts are interesting, in an academic sense—a fox hunt is where a transmitter is hidden and you have to discover its location using a portable radio set, as an exercise is learning how radio waves transmit, bounce, and generally are somewhat finicky. My context is that I only want to talk to machines. AMSAT, again, is the subset of ham radio people who work on satellites. But their primary work is not satellites *per se*, but satellites as radio repeaters to extend the range and reach of their person-to-person communications.

There is also a mild generational issue. Ham radio predates IRC and the Internet. Once the Internet appeared, fewer people needed a free avenue to talk worldwide because the Internet supplanted the role ham used to serve. Modern ham radio is mostly made up of older generation participants, plus a few clueful hackers who realize it's a secret public-use area that anyone can play in.

For the antisocial, you don't need to join the club or even talk to them, you can just take the test. Within this book, we are using radio to only talk to machines. It's a different world, and some hams may be upset to find out you don't want to use your license for standard ham activities like talking to each other. The radio subgroup that is most involved in *talking with machines* is AMSAT. The bulk of AMSAT satellites are radio relays, existing to facilitate people talking to other people.

Ham radio is the use of amateur radio gear to transmit and receive signals on allocated frequency bands. Whether your project requires it or not, you should get your ham radio license. It is perhaps the single best way to quickly learn the fundamentals of radio communication and, equally important, radio licensing and legal requirements. Anyone can stick a radio onto a project, but using a radio accurately, legally, and with respect for others is best served by passing the Ham test.

In any case, you will need to get your Technician's License, the first level of amateur radio certification. This is required if you operate any transmitter. It's worth getting even if someone else on your team is handling comm, because the exam—and the question pool for studying—is an excellent crash course in essential radio comm. The test for the Technician's License (the first level of ham licensing) is well documented and on the Web. Printed study guides and free web practice exams are both available. The test itself is usually done by volunteers from your local ham club.

You did know you have a local ham club, yes? Most countries, cities and towns have a group of licensed ham users who both assist newcomers in getting their licenses, and run local events.

Ham Technician's License

If you are in the US, here is how to pass your ham radio Technician class license. You read up, then go to a testing center and pay a fee that ranges from $0-15, and take a 35-question multiple-choice test; then, if you've scored at least a 70%, you are licensed for limited operations in all VHF and UHF bands above 30 MHz, and some specific HF bands. This means you can to talk to satellites. The learning you gain by practicing for this exam is priceless.

If you've taken college physics, you can probably score a 60% just taking a practice test cold. About one-third is basic physics/electromagnetics, one-third is jargon, and one-third is policy and common sense. You can get very far just knowing that frequency * wavelength = c (the speed of light). That's the one fundamental equation you also need for this book, because it lets you translate between frequencies (such as 440 MHz) and wavelengths (such as 70 cm).

From the National Association for Amateur Radio (ARRL) "LEVEL 1: Technician Class License," taken verbatim:

> The FCC Technician License exam covers basic regulations, operating practices and electronics theory, with a focus on VHF and UHF applications. Morse code is not required for this license. With a Technician Class license, you will have all ham radio privileges above 30 MHz. These privileges include the very popular 2-meter band. Many Technician licensees enjoy using small (2 meter) hand-held radios to stay in touch with other hams in their area. Technicians may operate FM voice, digital packet (computers), television, single-sideband voice and several other interesting modes. You can even make international radio contacts via satellites, using relatively simple station equipment. Technician licensees now also have additional privileges on certain HF frequencies. Technicians may also operate on the 80, 40 and 15 meter bands using CW, and on the 10 meter band using CW, voice and digital modes.

They (the ARRL) also publish a helpful book. Their arrl.org website includes where classes are offered, where you can take the test, and the entire question pool of possible questions. My advice is read their site, buy a study guide, practice the test online, then take it. You only need a 70% to pass, but why settle for passing? Go for 100%. The more you know, the better your satellite comms will be.

After passing the "Technician" class test, go for your "General" class, then the "Extra" class. None of the three classes require that you learn Morse code, an older requirement long since dropped.

For the Technician test, the ten categories include:

1. FCC Rules
2. Operating Procedures

3. Radio Wave Characteristics
4. Amateur Radio Practices
5. Electrical Principles
6. Electrical Components
7. Station Equipment
8. Modulation Modes and Amateur Satellite Operation
9. Antennas
10. Antenna safety

Sample Technician Questions

Here is one sample question for each category, taken verbatim from hamexam.org:

```
T1A04 (C)
Which of the following meets the FCC definition of harmful
interference?
A. Radio transmissions that annoy users of a repeater
B. Unwanted radio transmissions that cause costly harm to
radio station apparatus
C. That which seriously degrades, obstructs, or repeatedly
interrupts a radio communication service operating in
accordance with the Radio Regulations
D. Static from lightning storms

T2A04 (B)
What is an appropriate way to call another station on a
repeater if you know the other station's call sign?
A. Say break, break then say the station's call sign
B. Say the station's call sign then identify with your
call sign
C. Say CQ three times then the other station's call sign
D. Wait for the station to call CQ then answer it

T3B03 (C)
What are the two components of a radio wave?
A. AC and DC
B. Voltage and current
C. Electric and magnetic fields
D. Ionizing and non-ionizing radiation
```

T4A02 (D)
How might a computer be used as part of an amateur radio station?
A. For logging contacts and contact information
B. For sending and/or receiving CW
C. For generating and decoding digital signals
D. All of these choices are correct

T5C09 (A)
How much power is being used in a circuit when the applied voltage is 13.8 volts DC and the current is 10 amperes?
A. 138 watts
B. 0.7 watts
C. 23.8 watts
D. 3.8 watts

T6D01 (B)
Which of the following devices or circuits changes an alternating current into a varying direct current signal?
A. Transformer
B. Rectifier
C. Amplifier
D. Reflector

T7B04 (D)
Which of the following is a way to reduce or eliminate interference by an amateur transmitter to a nearby telephone?
A. Put a filter on the amateur transmitter
B. Reduce the microphone gain
C. Reduce the SWR on the transmitter transmission line
D. Put a RF filter on the telephone

T8B01 (D)
Who may be the control operator of a station communicating through an amateur satellite or space station?
A. Only an Amateur Extra Class operator
B. A General Class licensee or higher licensee who has a satellite operator certification
C. Only an Amateur Extra Class operator who is also an AMSAT member
D. Any amateur whose license privileges allow them to transmit on the satellite uplink frequency

T9A09 (C)
What is the approximate length, in inches, of a 6 meter

```
1/2-wave length wire dipole antenna?
A. 6
B. 50
C. 112
D. 236

T0B09 (C)
Why should you avoid attaching an antenna to a utility
pole?
A. The antenna will not work properly because of induced
voltages
B. The utility company will charge you an extra monthly fee
C. The antenna could contact high-voltage power wires
D. All of these choices are correct
```

Scenarios

One area I wish to emphasize is that you have to plan out your command and control scenarios. It is not enough to say, "I have a radio and I know how to use it." Key issues are:

- Number of ground stations + locations + orbit = # of potential contacts per day
- Duration of contacts = 1) amount of data you can download + 2) number of command opportunities

For most missions, real-time control is infeasible. You will typically spend one pass downloading data and uploading commands for the next few orbits. "Joysticking" your satellite to act in real time is a luxury for... well, no one. It's a bad idea. Big NASA Missions don't let someone twist a multimillion dollar mission on a whim. Even your small satellite should have plans, scenarios, and checklists for carrying out expected operations.

Safehold

The Safehold is your friend. Safeholds (or Safing) means that your satellite experienced something that put it into an unknown situation. So, the satellite instead put itself into a minimum operational mode. That mode usually says:

- Power down everything nonessential

- Maintain communications
- Do as little as possible until diagnosis from the ground tells it what to do

The wrong response to satellite trouble is, "Let's try this." The proper response is to analyze what may have caused things, and test situations in a ground simulator until you have a working solution.

Ground Sim

If you can't communicate with your satellite across a 10-foot lab bench, you'll never succeed in orbit. So build a *flatsat* (tabletop satellite hardware mock-up or simulator) and a rough ground station and begin testing.

Hacked-up mobile ground station: wire a walkie-talkie or SDR USB plug to a laptop and grab an antenna and you're ready to go.

Software

We have talked at an organizational and functional level about the components that need to interact to produce a successful space mission. At a software/network/interface layer, how do they talk? The terms *architectures*, *frameworks*, and *buses* are used to define the system. You define at a high level what your mission needs and assess what your organizational elements (or functionalities) are.

At the implementation level, you choose an underlying bus or protocol or standard for building your mission operations components (e.g., GMSEC, CCSDS, etc.). The bus connects all of your organization elements.

Then, to do actual operations, you use or create modules that carry out the actual functions or work.

Compare and contrast

GMSEC bus
: An underlying bus where the message standard between components is defined

CCSRS SOA
: A service-oriented architecture (SOA), where a metadata standard is defined but the components and protocols are not specified

ITOS
: A GSFC command and telemetry system that includes a simulator

CFS
: An environment library based on layered architecture and open standards

For example, XML is an SOA, http: is a bus, WordPress is a set of tools, and Ruby on Rails is an environment.

Laser Comm?

Using lasers to communicate, instead of radio, is relatively new. The Lunar Atmosphere Dust Environment Explorer (LADEE) mission (*http://bit.ly/1K64SfA*) tested a 20+Mbps laser communications system across the Earth-Moon distance. At 32 kg and 137 watts of power, the satellite laser comm unit sends data to a ground station consisting of eight optical telescopes.

Laser communications is to a large degree not regulated by the FCC (which handle radio), but by the FAA (which ensures aircraft are not at risk from laser light). In theory, laser can give point-to-point communications with little spillover or risk of being intercepted by someone out of the line of sight, and promises to deliver high bandwidth rates. The equipment isn't there yet, but that means this is the type of mission picosatellites can test.

Relays

Getting signal from the spacecraft to the ground can be by direct transmission or via a communications relay. The following demonstrates line of sight versus relay and store and forward:

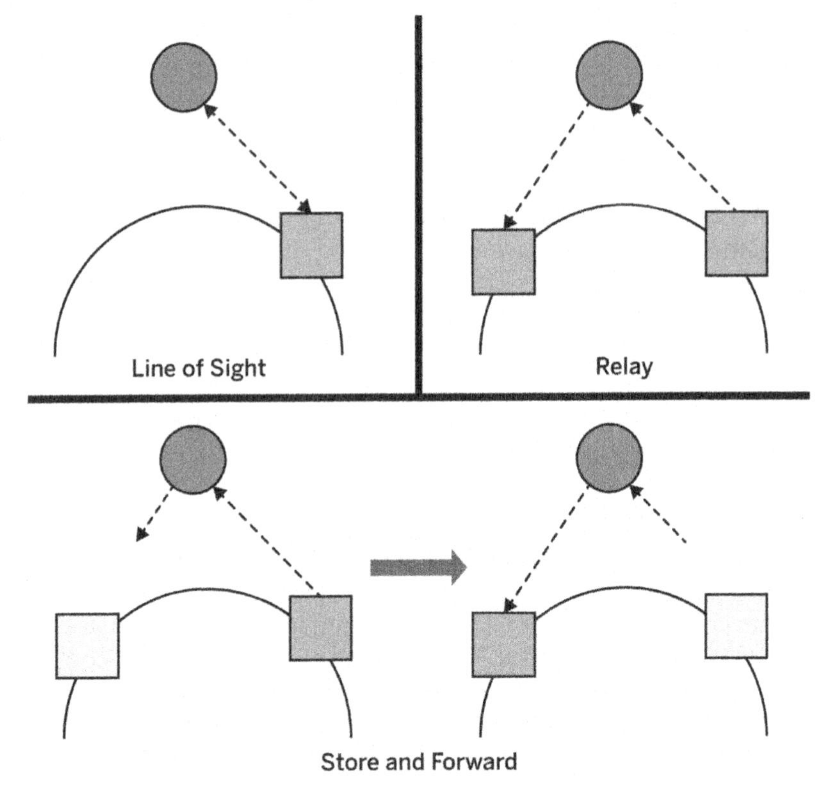

Line of Sight/Direct to Ground

Direct to ground requires the satellite to be over the horizon and within line of sight of the chosen ground station. The term *line of sight* means exactly what it says, being able to see the satellite. The biggest obstruction is the Earth, hence the requirement that the satellite be over the horizon. Simply put, if you could aim a telescope at your satellite and see it, you can reach it with a radio.

One reason communications stations are often up on mountains is that they provide more *horizon* with fewer obstructions at low angles. Buildings block (or weaken, or *attenuate*) radio signals, as do trees, bridges, or just about any chunk of matter between you and the satellite.

Relays

A *relay* is a radio transmitter/receiver used to increase the range of a transmission. It receives a presumably weak signal, then re-broadcasts it at a higher strength. Like a relay race, a series of repeaters can receive then retransmit across long distances. A repeater can immediately retransmit, or use a mode called store-and-forward, in which the signal is held for a bit and transmitted later. The bulk of AMSAT satellites are repeaters intended to relay amateur radio communications to a distance further than the individual user could usually reach.

Relays can instantly retransmit, or can use a store-and-forward mode where they store incoming messages, then retransmit once a given criteria is met. An automatic repeater that instantly retransmits assumes that there is a receiver ready to catch. A store-and-forward system, in contrast, waits until a receiver provides notification that it's ready to receive, then sends the message. For example, a satellite might receive a message while overhead Canada, intended for Hungary, and then store it and wait until it *sees* a Hungary ground station before forwarding it.

Two examples of a repeater network for communications are the TRDRSS geosynchronous communications satellite, and the Iridium satellite constellation. One design is a permanently floating relay station with a 24/7 connection to its single ground station, and the other is a swarm of moving satellites that pass messages satellite to satellite until they see a ground station.

An individual TDRSS satellite, from the ground, seems to always be in the same position of the sky, and can be pointed to with a fixed antenna. It receives signals from multiple satellites in the clear of space, where no atmosphere impedes the signal, then retransmits those signals to a ground station in White Sands, who then reroute the signals through conventional Internet to users.

The Iridium satellites, in contrast, are 100+ satellites all moving around the Earth every 100 minutes, evenly spaced. Your signal goes to the nearest satellite for which you have line of sight. That satellite then retransmits your message to the nearest Iridium satellite it can see, which transmits it to the nearest Iridium *it* can see. And so the message is passed along the network of moving satellites until it finally reaches a satellite that happens to have line of sight to the destination; that satellite transmits the message to ground.

Building a geosynchronous relay satellite like TDRSS specifically for amateur, CubeSat, and university-class satellite use is a project several groups are tackling. Similarly, a cheaper Iridium-like constellation of CubeSats acting as a comm relay is a project two other groups are tackling.

Still, others are taking a regulatory stance and hoping to get permission to open up part of the amateur spectrum to be either:

1. Dedicated to amateur satellite use and/or
2. Allowed to use encoding and methods that increase the amount of data transmitted

Currently, amateur satellites have to transmit in the clear, such that anyone can receive it, as the amateur bands are a shared resource for participation by all. Using encoding that requires special hardware or software (to boost throughput) is not yet allowed.

Getting onto a Network

Just because space-based comm relays exist is no guarantee you can use them. Use of a TDRSS relay requires paying for access, and purchasing/having the specific transceiver they require. Use of Iridium for space is not yet licensed, though a group is prototyping this.

Cell phones use short-range towers because the typical cell phone signal is weak and fairly easily attenuated, and thus doesn't travel far. However, in space, even a weak signal has a greater range. It just needs someone to agree to send it to the ground. There is no reason existing smartphone and satellite phone networks could not work in space. Radio is radio, and the only difference between communicating from your house or from space is the Doppler shift your fast-moving satellite has. This is a mild implementation problem for repurposing a non-space-use network for space, but the Doppler calculation is both known and solved. So it's not a technological barrier, but is an implementation issue.

Licensing is very serious. For example, a high-altitude balloon can go up 70,000 feet or more. A student project (Hermes) has flown an Iridium satellite phone side by side with a ham radio to measure performance. As the balloon went up, the Iridium device cheerfully sent data until it hit 10,000 feet, when it sent one final message: "Your device is not licensed for air use, please contact your sales agents." It was not a hardware limitation, but a licensing one, that made the device stop working.

Private Networks

The luxury car of satellite communications is paying a commercial provider to carry your traffic. This solution says to make comm "someone else's problem." You send your commands via the Internet to the provider, they handle the uplink to the spacecraft, and later they provide the downlink of your data to your ground computer. Highly recommended, this is the growth area for CubeSats in 2014-2016. Some existing networks include:

- DSN = for LEO, 5 passes of 10+ min each from the 15 orbits/day.
- TDRSS = potentially 24/7 coverage via NASA network.
- Private networks offer a mix of these.

Sample Missions

We provide several writeups of existing missions, in order to better present how others have implemented their specific communications architecture. Most CubeSats send *beacon* messages, which are short messages including a little HK data and primarily exist to show the satellite is alive. Actual mission data or images are then sent separately:

SkyCube

Beacons contain 120-byte user-submitted messages. Images are also sent. Both use AX.25, 9600 baud, BPSK modulation on 915 MHz. Uplink is at AX.25 packets, 9600 baud, FSK modulations on 450 MHz. Call sign is WG9XM.

ESTCube-1

Beacon data sent using telegraph signal on 437.250 MHz. Data is sent with FSK modulation on 437.505 MHz. Both use AX.25. Higher rates are available for data using GFSK-, MSK- and GMSK-modulations.

COMPASS-1

Morse code beacon every three minutes is on 437.275 MHz. Data is AX.25 at 1200 baud (AFSK) or 2400/4800 baud (MSK) on 437.405 MHz. Uplink is DTMF tones on 145.980—yes, *tones* means it commands by sending specific tones, much as an old push-button phone would send a specific tone to pass along dialing information.

FUNCube

Beacon sends at 145.935 MHz. Data sent as AX.25 at 1200 baud with BPSK on 145.950-145.970 MHz, including an unmodulated BPSK tone so you can literally hear—and thus identify—that this is FUNCube data. Uplink is 435.150-435.130 MHz.

5/Orbits

Defining Orbits

It's helpful to know where your satellite is, other than saying "up there" or "in orbit." For Earth-orbiting satellites, their path is a precise, repeating ellipse with one focus at the center of the Earth. An orbit is defined by six parameters, and if you know the value of those parameters, and the time, you can predict exactly where your satellite will be.

A Little Theory

The current movement of anything (orbit or not) can be defined by two terms: its position and its velocity. We live in a three-dimensional world, so we need three coordinates to define its position as x, y, and z. We also need three coordinates to define its velocity—velocity along x, along y, and along z. That's six coordinates to define our item.

Those six parameters can be either the state vector, or the classic orbital elements. We call the position coordinates vector *r* (x,y,z) and the velocity coordinates vector *v* (vx, vy, vz). With *r* and *v* you have the instantaneous state of the object. So in three-dimensional space, six data points for our state vector (r,v) fully defines our satellite—for that one moment in time. If the satellite moves purely under the influence of gravity (no rocket firings allowed), you can also predict its future position.

Kepler worked out that orbits move in a geometric pattern called an ellipse. Newton later co-invented calculus to define why, but you don't need to use that. You just need to know it's a solvable problem. If you know the state vector (r,v) at some initial time, you can predict exactly where the satellite will be at any time in the future.

To solve this yourself, you can either take an orbital mechanics class or—as we recommend—you can use software. Orbit prediction codes are available freely on the Internet, and there are web apps that solve the calculation for you. The "Where Is the ISS?" links in Chapter 7, for example, use this kind of software to generate their prediction.

Don't Go It Alone

This leads to one rule for ground operations—do not write your own code! There is code out there for sending and receiving data, for orbital predictions, for operating the antenna. You do not have to invent anything new to operate a satellite. The ground and comm segments are an integration task, where you have to understand the components well enough to carry out your plan.

The state vector (r,v) is only one way to calculate an orbit. There are other sets of coordinates that can be used. One such set is the classic orbital elements, which define the orbit in terms that allow for easy prediction, and are the data shared using two-line element (TLE) data sets.

The Six Classic Orbital Elements

The six classic elements include terms that describe the shape of an orbit (semi-major axis *a* and eccentricity *e*), terms that define its position relative to the Earth's equator and poles (inclination *i* and argument of perigee *w*), a term that ties the orbit to our calendar (longitude of the ascending node *O*), and a final item that says, for the defined orbit, how far along it is (usually *mean anomaly*, but also can be *true anomaly* or *angle from last perigee passage* or *time from last perigee passage* or *time of periapsis passage*).

1. a = semi-major axis = size
2. e = eccentricity = shape
3. i = inclination = tilt
4. ω = argument of perigee = twist

5. Ω = longitude of the ascending node = spin
6. v = mean anomaly = where in orbit

Shown here is "Orbit1" by Lasunncty (talk) (licensed under CC BY-SA 3.0 via Wikimedia Commons):

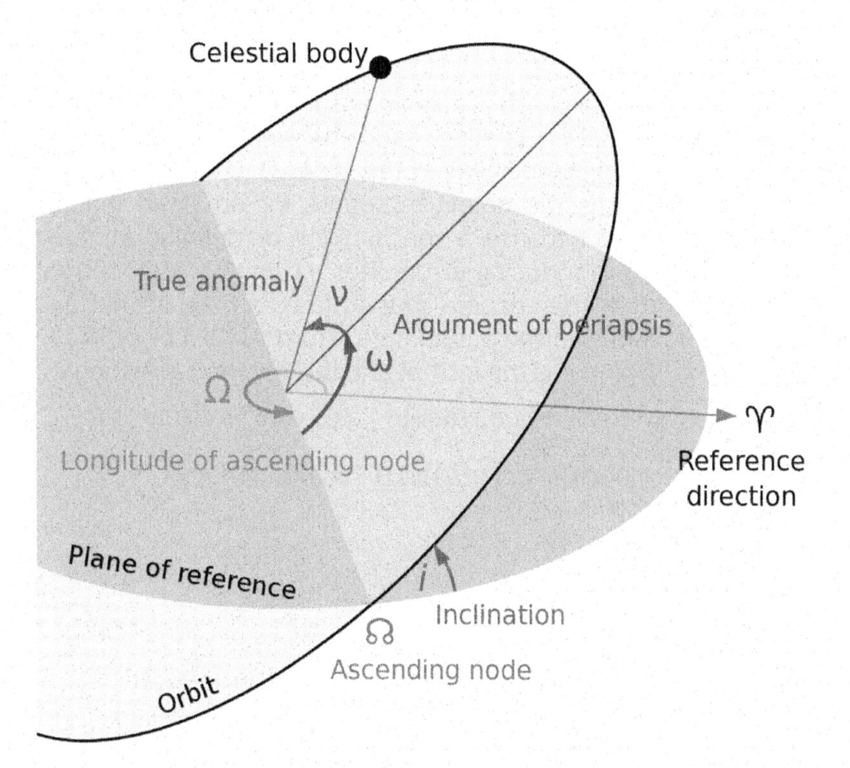

It is not so important that you understand the full meaning of these elements (you can read more online to learn that), but that you accept that these six elements fully define an orbit and its current position.

You can convert any state vector (r,v) to its equivalent orbital elements, and you can convert any orbital elements to a state vector. The state vector says, "Here is a direct line to the object, now," and the orbital elements are a tool that lets you predict any future state vector. They work well together.

One neat property of the orbital elements is that an orbit's period *P* (time to complete one full 360 lap) is proportional to its semi-major axis *a* (its size). $p^2 = k\ a^3$, where the constant *k* depends on the units used for *a* and *p* (and the mass of the object being orbited around; k is different for orbiting Jupiter versus orbiting the Earth, for example). This means that if you know *a*, you also know the period of the orbit—and vice versa.

TLEs

Orbital elements are often exchanged in NASA's TLE format, covered in Book 1, DIY Satellite Platforms. Here is a short definition of the specification, excerpted from AMSAT.org, in part because AMSAT is the definitive source for amateur satellite work, and in part because a specification or definition is best quoted to avoid introducing errors. It is a primarily intended for computers rather than people, in that it has a fixed format and there are no labels in the actual data. From AMSAT.org (*http://bit.ly/1K67UjO*), here is the format and the key for translating it:

```
1 AAAAAU YYLLLPPP  BBBBB.BBBBBBBB  .CCCCCCCC  DDDDD-D  EEEEE-E
F GGGGZ
2 AAAAA  HHH.HHHH III.IIII JJJJJJJ KKK.KKKK MMM.MMMM
NN.NNNNNNNNRRRRRZ

Key

[1] - Line #1 label
[2] - Line #2 label
[AAAAA] - Catalog Number assigned sequentially (5-digit
integer from 1 to 99999)
[U] - Security Classification (U = Unclassified)
[YYLLLPPP] - International Designator (YY = 2-digit Launch
Year; LLL = 3-digit Sequential Launch of the Year; PPP = up to
3 letter Sequential Piece ID for that launch)
[BBBBB.BBBBBBBB] - Epoch Time--2-digit year, followed by 3-
digit sequential day of the year, followed by the time
represented as the fractional portion of one day
[.CCCCCCCC] - ndot/2 Drag Parameter (rev/day2)--one half the
first time derivative of the mean motion. This drag term is
used by the SGP orbit propagator.
[DDDDD-D] - n double dot/6 Drag Parameter (rev/day3)--one
sixth the second time derivative of the mean motion. The "-D"
is the tens exponent (10-D). This drag term is used by the SGP
orbit propagator.
[EEEEE-E] - Bstar Drag Parameter (1/Earth Radii)--Pseudo
```

```
Ballistic Coefficient. The "-E" is the tens exponent (10-E).
This drag term is used by the SGP4 orbit propagator.
[F] - Ephemeris Type--1-digit integer (zero value uses SGP or
SGP4 as provided in the Project Spacetrack report.
[GGGG] - Element Set Number assigned sequentially (up to a 4-
digit integer from 1 to 9999). This number recycles back to
"1" on the update following element set number "9999."
[HHH.HHHH] - Orbital Inclination (from 0 to 180 degrees).
[III.IIII] - Right Ascension of the Ascending Node (from 0 to
360 degrees).
[JJJJJJJ] - Orbital Eccentricity--there is an implied leading
decimal point (between 0.0 and 1.0).
[KKK.KKKK] - Argument of Perigee (from 0 to 360 degrees).
[MMM.MMMM] - Mean Anomaly (from 0 to 360 degrees).
[NN.NNNNNNNN] - Mean Motion (revolutions per day).
[RRRRR] - Revolution Number (up to a 5-digit integer from 1 to
99999). This number recycles following revolution number 99999.
[Z] - Check Sum (1-digit integer). Both lines have a check sum
that is computed from the sum of all integer characters on
that line plus a "1" for each negative (-) sign on that line.
The check sum is the modulo-10 (or ones digit) of the sum of
the digits and negative signs.
```

As an example, here is a TLE for the X-Ray Timing Explorer (XTE) satellite telescope (now long since deorbited). I include the common (but optional) extra first line designating the mission name (*XTE*):

```
XTE
1 23757U 95074A 11333.55435324 .00010246 00000-0 30421-3 0 5229
2 23757 022.9893 121.5427 0007734 071.5825 288.5466
15.321190098829520
```

Not human-readable, but that is not the point. TLEs are designed so software can easily read the fixed-format information and operate on it. As long as you adhere to the standard, you can get TLEs from any authoritative source and your orbit prediction software will be able to use it. You can get TLEs for just about every tracked satellite from Space-Track (*https://www.space-track.org*), which is chartered by the US Department of Defense to distribute TLEs.

Orbit Determination

So where do TLEs come from? Most people will wait until NORAD/US Space Command releases their orbit data. However, there are five common ways to determine, from observations, the orbital elements. The first method is direct ranging using radar, which is very handy if you own your own radar installation but usually out of the realm of the amateur. The other four are mathematical methods where you make multiple observations of the satellite during one orbit, then use the appropriate mathematics to derive the orbital parameters. These four methods are triangulation, Gauss's Method, Lambert's Problem, and the Gibbs Method.

Direct ranging using radar (with Doppler shift), gives you the state vector (r,v) at some known time *t*. With this, you can derive the orbit because each orbit is unique and independent of the mass of the spacecraft. As mentioned back in Book 1, *DIY Satellite Platforms*, a satellite's orbit does not depend on its mass, only its position and velocity. Both the ISS and a CubeSat next to it will orbit identically. This is why, when they eject CubeSats from the ISS, they give the CubeSats a little velocity kick—so they won't remain right next to the ISS.

Using the radar timing to derive the distance plus the Doppler shift on the radar to get the velocity, you obtain a position and velocity of the satellite relative to the radar station. Add some geometry based on your latitude and longitude, and you obtain the exact state vector (r, v) for the satellite. This is beyond most non-radar-owning amateurs, but is the method our various governments use to track space items. If you don't have this, you can use a series of ground observations to deduce the orbit. Which method you use depends on that data you can gather.

The remaining methods use multiple ground observations of the satellite during one orbital path. If, from the ground, you can observe the altitude (height above horizon) and azimuth (compass angle), or alternately the astronomical right ascension and declination, of a satellite, you know it is somewhere along that line of sight but you don't know how far it is.

Triangulation consists of using two simultaneous measurements of an object from two different ground locations to geometrically calculate the object's exact position. Each observation is one side of a triangle; the line connecting your two observers is the third side. Knowing these three sides and the angles involved, you can use triangle geometry to get the exact position. Therefore, two observers can get a single *r* vector for a satellite, if they are well coordinated in gathering their data. Satellites zip across the sky quickly, and triangulation requires the measurements be at the exact same time, but this is doable.

If you take three different observations of the object, you can use Gauss's Method to deduce the complete orbital elements. This is the oldest and roughest method. Code to solve Gauss's Method exists online and in orbital mechanics textbooks.

If you use radar or communications ranging or triangulation to get the distance, you have a position vector. With two exact position vectors (direction plus the distance to it) and the flight time between them, Lambert's Problem is an algorithm that will give you the complete orbital elements. With three position vectors but no time information on when the data was taken, you can use the Gibbs Method (implemented in software) to derive the velocity vector *v* and fully define its state vector. Again, code to solve both is easily found. You now know its orbit.

Time

Satellite work uses Universal Time (UT) and Julian Date (JD) or Modified Julian Data (MJD). These are absolute time standards that do not consider time zones, local time, daylight savings time, or other human-centric factors. We use them for satellite work so we can share data using the same standard.

JD is the number of days since noon (UT) on January 1, 4713 BC. Yes, this tends to be a big number. MJD is a shorter version that also has the advantage of starting at midnight instead of noon UT. The formula is easily found online or in any astronomy textbook and is given, for a desired day/month/year, as shown here (US Navy, Fliegel and van Flandern algorithm, 1968 (*http://1.usa.gov/1K698vs*)):

```
JD = Day-32075+1461*(Year+4800+(Month-14)/12)/4+367*(Month-2-
(Month-14)/12*12)/12-3*((Year+4900+(Month-14)/12)/100)/4
```

This formula works best with a strongly typed computer language like Java or C, not with "duck typed" languages like Python. It assumes integer math for some of the leap day/year calculations. See the original source URL for specifics.

And the handy Modified Julian Day, MJD, is just: *MJD = JD – 2400000.5*.

Universal Time (UT) is simply *midnight at Greenwich, England*, which is midnight at the 0 latitude line of the Earth. UT runs as a 24-hour clock. You can convert from UT to your time zone by knowing your offset from UT. UT is sometimes considered the same as Greenwich Mean Time, which is sometimes (especially in the military) called Zulu.

Time can get finicky pretty quickly. There are slight variances between atomic time (UTC) and astronomically-defined time (UT1), and occasionally leap seconds are added to a year to align the two. For satellite use, fortunately, if you just stick with UT and MJD, you can freely exchange data and rely on software without having to delve into the details of time.

Satellite Time

Your satellite will probably not know what time it is. It doesn't have to. You can imagine your satellite is in its own time zone. It is not important what time is on its internal clock, only that you know the offset between your satellite time and terrestrial UT.

When your satellite first powers up, it will have a default *start* time on its internal clock. This is the equivalent of a newly plugged in alarm clock reading *00:00*. You need to record this time and compare it to your ground time. This gives you your clock offset.

While you could reprogram the satellite clock to the *current* time, this is not recommended. First, it is hard to do this accurately, and second, you have to know exactly when you sent the command plus how long it took the satellite to react to the command and reprogram. It's also totally unnecessary. Instead, just let the satellite keep its own time, and remember its offset.

For example, if your spacecraft powers up at 17:57UT on Nov 14, 2014, but thinks it started at 12:00 Nov 1, 2014, you now know its offset is that the satellite is running 14 days and 5:57 hours behind your *real* time. Keep that offset and don't mess with the satellite clock, because it's unnecessary to do so and brings absolutely no advantage. Just let the clocks run as they are.

Relativity does alter time. Two objects moving at different speeds or facing a different amount of gravity will find out their clocks run at different rates. The two relativity factors (velocity and gravity) operate in opposite directions, so it turns out the ISS clocks run slower than on Earth (high velocity, roughly the same gravity), while GPS clocks at geosync orbits run faster (slower, further from gravity). The correction can be predicted, but it is easier to just periodically check the clocks against each other and update the offset.

The one implication for commanding your satellite is that if you have any timed commands you wish to do (anything programmed other than *do this as soon as you receive this transmission*), you need to make your command load match the spacecraft clock/date, not your ground date.

Most larger missions will, in their spacecraft software, include a data item of this offset so the spacecraft is aware of the time shift. For CubeSats, which typically run smaller processors with more immediate "do this as soon as told" commanding, this may not be necessary.

Reboots are another factor to consider. If your satellite reboots itself, either on command or due to radiation damage or a safehold event, the spacecraft clock might (yet again) reset itself to a new time. Just update your offset and you are good to go.

Humor: "Give Me an Orbit"

"We need a transfer orbit from LEO to GEO!" Orbital mechanics for...

Nontechies
: Why can't they just point and go straight there, like in the movies?

Scientist
: Assume a Hohmann transfer, a quick napkin calc shows our two fjvs and time. Done!

Engineer
: Give me the start and end points you want and I'll run it through the simulator.

CS & Math
: I think our code will converge quicker if we use Newton-Rapheson instead of Bisection.

Management
: We can give you a pork chop plot of solutions today, but actual firing data will have to come from FDF.

Ground Tracks

The final part of the picture is to project the elliptical orbit of the satellite onto the Earth's surface, so you can see (at any given time) what spot on the Earth your satellite is over. This projection is the ground track. Ground tracks reveal an interesting property of orbits. While the orbit is fixed in space (nothing to alter it) and keeps tracing an eternal ellipse around the Earth, the Earth is meanwhile rotating beneath it. An orbit once around the Earth travels over a *swath* of land, but because the Earth rotated in that time, for most orbits, the start of the next orbit does not hover over the same spot as the start of the previous.

It is important to understand a ground track, because at a glance it tells you a lot about a mission. It gives you a rough idea of its orbit (low earth, geosynchronous, other), it tells you the orbital period, and it visualizes the times and places the satellite can see and can communicate with the ground. Here is a sample ground track showing two orbits in a row for the ISS (licensed under public domain via Wikipedia):

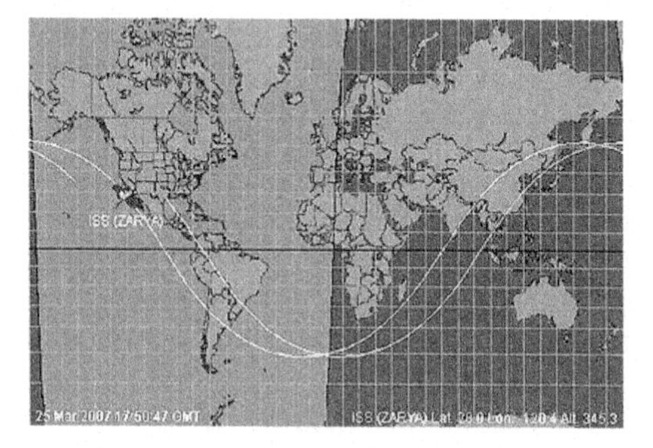

Each orbit of ISS takes about 100 minutes. During those 100 minutes, the Earth rotates 25 degrees in its daily cycle. So each ISS orbit ends up about 25 degrees apart from each other, in longitude.

In fact, from a ground track, you can deduce an orbital period. Here is a way of saying, "During one orbit, what fraction of the Earth's 24-hour day did it rotate?" Take the longitude spacing between two equatorial crossings that go in the same direction (i.e., crossing the equator while heading north), in degrees, divide by 360 degrees, and multiply by one day: $Period = 24hours \cdot (Spacing)/360$.

A second item is that you can tell the inclination *i* of the orbit by seeing what the highest (or lowest) latitude reached by the orbit is. That's its inclination. In the case pictured, it's roughly at 50 degrees of inclination.

Since you can derive the period (and therefore the semi-major axis) and the inclination from the ground track, that means the ground track provides two of the six orbital elements you need, without any advanced math.

There are many kinds of orbits, but the most common for CubeSats are low Earth orbits. These tend to be in the 250-500 km altitude range, with periods on the order of 90 minutes per orbit (depending on altitude and eccentricity). Eccentricities vary greatly—some are very eccentric (very elliptical), while others have nearly no eccentricity (circular). Software (of course) exists for plotting ground tracks given the orbital TLEs.

Every ground station should have a constantly updating plot of the ground track, not just for coolness but because it gives you a quick visual sense of when the satellite will next be available overhead to your ground station.

You will often find the requirements for your ground station broken down into three parameters—coverage area, slant range, and viewing time. Coverage area is, at any given time, the area on the ground that can *see* the satellite. Slant range is the line-of-sight distance from ground antenna to spacecraft. Viewing time is the duration that the ground station can *see* the spacecraft, assuming it is a suitable distance above the horizon.

Contact Passes

Your operational flow sequence is to build a schedule for contacts, consisting of:

1. Acquisition of satellite (AOS, also called acquisition of signal)—the time the satellite is above the horizon and expected treeline, and thus visible to your station
2. Location at AOS (relative to your station)
3. Loss of satellite/signal (LOS)—the time the satellite drops below the treeline and/or horizon and is no longer visible
4. Location of LOS
5. Path of satellite

This is easier done in software. Briefly, orbits by definition are a fully predictive path. The most common reference frame for the orbit coordinates is the geocentric frame. Software is used to map this to your local position on the Earth (the topocentric frame), and thus translate the satellite's position into either an azimuth (compass direction to face) and altitude (height above the horizon), or a right ascension, declination pair (sky coordinates to point at), that your system can then track.

From the predictive elements you can visualize the ground tracks, and also generate a handy table of when your satellite is visible for a specific ground station. Here's a sample set of data for the ISS for a typical day, generated by SATFLARE (*http://www.satflare.com*). For December 18th and 19th, there are seven consecutive passes per day ranging from two to five minutes. If you rule out passes below five minutes, you get four *good* passes per day. The following shows a sample set of data for the ISS on a typical day:

Rise	Peak	Set	Best Time	Mag.	Sat El. at Best	Sun El. at Best	Visible
Thu, 18 Dec 2014 12:23:06	12:25:22	12:27:39	12:25:22	Shadow	1.9	27.5	No
Thu, 18 Dec 2014 13:55:54	14:01:10	14:05:26	14:01:10	Shadow	40.2	22	No
Thu, 18 Dec 2014 15:32:44	15:3750	15:42:57	15:37:50	Shadow	24.3	10.2	No
Thu, 18 Dec 2014 17:11:03	17:15:22	17:19:41	17:16:22	-0.1	8.9	-5.7	Twilight
Thu, 18 Dec 2014 18:48:43	18:53:11	18:57:39	18:50:57	0.7	7	-23.1	Low Elevation
Thu, 18 Dec 2014 20:25:15	20:30:30	20:35:45	20:30:30	Shadow	36.4	-42.4	No
Thu, 18 Dec 2014 22:01:59	22:06:56	22:11:53	22:06:56	Shadow	21.8	-60.6	No
Fri, 19 Dec 2014 13:05:39	13:10:37	13:15:35	13:10:37	Shadow	21.1	25.8	No
Fri, 19 Dec 2014 14:41:45	14:47:02	14:52:20	14:47:02	Shadow	38.2	17.1	No
Fri, 19 Dec 2014 16:19:50	16:24:20	16:28:51	16:24:20	Shadow	11.6	3.1	No
Fri, 19 Dec 2014 17:57:51	18:02:09	18:06:27	18:02:17	-0.1	9.7	-13.9	Low Elevation
Fri, 19 Dec 2014 19:34:37	19:39:40	19:44:44	19:35:02	1.9	1.5	-31.5	Low Elevation
Fri, 19 Dec 2014 21:11:07	21:16:21	21:21:36	21:16:21	Shadow	41.9	-51.1	No
Fri, 19 Dec 2014 22:49:49	22:52:10	22:54:32	22:52:10	Shadow	2.1	-68.1	No

6/Comm Budgets

Allocations

If you are a satellite builder, you are going to have some sort of bandwidth limit on how much data you can send and receive from your spacecraft. To fit your mission in that communications budget, you have to make a comm budget. Even if you aren't a communications engineer, even if someone else is providing your communications network and antenna and ground commanding and data handling for you, as the mission designer, *you* need to work out a comm budget.

A comm budget is, at its heart, something most people already do with their cell phone plans. Say you have an old plan that gives you 50 free text messages a month. That's the core of your comm budget—50 texts per month. What makes satellite comm budgets different than cell phones or Internet connections is that the consumer communications usually let you exceed your allocation, and just charge you more money. By analogy, if you go over your 50 free texts, you then pay $0.25 per text—but your texts still get delivered.

With a satellite, when you use up your communications allocation, you're done. Even if you don't finish your task, it's over. Therefore, you must predict your usage, and keep to your resources.

Your comm budget involves tabulating how much data you are sending up to the satellite and how much data you are receiving down, and comparing it to how much communications time you have available.

First, you allocate how many uplinks of commands you need, and how many downlinks of spacecraft data you require. Your downlinks will include both housekeeping (HK) health-and-safety data and actual instrument/payload data.

Housekeeping data is information on items like satellite power levels, temperatures, voltages, instrument diagnostics, and other items indicating how well the satellite is operating. HK may include position information as well, particularly if you are using a GPS to record the actual position. The amount of HK data is usually small, since it's the equivalent of a short summary chart or table of numbers.

Instrument data is what most people think of when you use the term *data*. It is the actual pictures or instrument readings or radiation levels or whatever your useful sensors (from Book 3, *DIY Instruments for Amateur Space*) are doing. This will usually be the largest part of your comm budget.

A contact pass is defined as "the satellite is visible and available to a specific ground antenna below." Contact passes can be just for listening, or include talking to or commanding a satellite.

Data Priorities

Different data items have different priorities or urgencies. Even though it's small, the HK data should be high priority because it tells you whether your satellite is alive or not. Similarly, the commanding is usually small in size but it a high priority because it tells the satellite what it needs to do.

The payload (or instrument) data is the largest in size, but the lowest in priority. You obviously want it, because it's the reason you flew a satellite—to carry a useful payload—but getting 100% of your mission data has to take second priority to making sure first that the mission is actually operating.

Put another way, if the satellite is OK and being commanded, you can suffer a little data loss and still have partial mission success. If you instead download data but the satellite itself is failing or doesn't know what to do next, you will acquire zero future data and your mission is dead.

Now we will break up our comm budget into components, and design our usage—how much comm we use for uplink, how much for HK downlink, how much for payload data downlink, etc.

Extended Example of Command via Text Messaging

Let's get back to our analogy, 50 texts per month. Assume that's a hard limit, with no overage or extra allocation available. Now say you are using something called *chirps* as your text messages—we are defining these imaginary tweet-like items as "140-character messages." We will use these *chirps* to communicate with your picosatellite. This is a concept actually being investigated by Rishabh Maharaja and Aaron Bush at Capitol College (thanks to Aaron for helping me work out this analogy). What does your comm budget look like?

You have 50 *packets* (the chirps) that each can contain up to 140 characters. We'll assume your *chirps* include a time tag already, so you don't need to add that to the data. All data overhead is being absorbed by your comm provider, as is encoding, so your only concern as the user is that you get 40 of these chirps as usable data transfer. We'll also ignore a lot of things we discuss later in this book involving security, authentication, and data integrity.

We'll make a comm budget for two possible satellites: "ThermoSat," which measures temperature fluctuations in the 3K environment of space, and "PhotoSat," which takes a 512 x 512 color image of the Earth. These missions need different amounts of data.

ThermoSat measures temperature. A temperature measurement is just a number, like 0 degrees or 300 degrees. So in one 160-character message, given a temperature (in degrees Kelvin) that will probably never be more than three digits in size (under 999 degrees C), you can fit (140/4) or 35 temperature readings per chirp. If you take one reading during each 90-minute orbit, a single chirp can contain 35 orbits and thus two days worth of payload data.

PhotoSat takes a picture, a 512 x 512 color image require three numbers per pixel and (again, as per Book 3, *DIY Instruments for Amateur Space*) if you record each color as a number from 0-255 and use ordinary text to represent it, your image-via-chirp will take up 3,145,728 characters. That means you need just under 22,500 chirp-like messages to convey a single photograph! So in this low bandwidth situation of 40 chirps per month, you simply do not have the bandwidth to transmit images.

Our first takeaway from comm budgets, then, is to realize when a given payload cannot fit within your budget. If your comm system has a low bandwidth of 50 chirps/month, you cannot fly a camera. Sorry, no go. Even if you want to get tricky and start using compression or special encoding, you're not going to get a 20,000x reduction in size. The solution is ditch the camera, or get a new communications provider or system that allows more bandwidth.

Having ruled out PhotoSat, we can explore ThermoSat within our given comm budget. We worked out one downlink chirp budget can contain 2 days worth of one-per-orbit data readings. So if we want to capture 100% of our data for a 30-day mission, that requires (30/2) = 15 of our chirps to acquire data.

Our comm budget narrative starts out as:

(v1) 15 passes per month, each pass sends 1 chirp down (1 chirp per pass, 15 chirp per month)

How will the satellite know when to send the data down, though? We have to command it to start a data downlink. So, each of those 15 downlinks will require we waste a chirp uplinking the order to the satellite of *begin transmitting*. So we are now using 2*15 = 30 of our 50 chirps per month to acquire the data.

This also means we want 15 communication contacts (or passes) per month with our satellite. Our comm budget, as a narrative, now reads:

(v2) 15 passes per month, each pass = 1 chirp up, 1 chirp down (2 chirps per pass, 30 chirps per month)

Because we also need to check the status of your satellite—the HK data—we want to add a downlink of HK data during our contacts. The HK data is just a snapshot of the system readings as the satellite currently is—no stored or past HK data to worry about—so we'll assume we can cram it all into one HK chirp. Assuming operationally we will get the HK at the same time as the data (during our 15 passes per month), we can use efficient commanding so the satellite knows that one command up tells it to send first the HK chirp then the payload data chirp. Therefore, we have to add 15 downlink chirps consisting purely of HK data to our budget.

(v3) 15 passes per month, each pass = 1 chirp up, 2 chirps down (3 chirps per pass, 45 chirps per month)

At 45 chirps/month, we are well within our 50/month comm budget, and each individual contact pass is neatly specified as 1 chirp up, 2 chirps down (first is HK, second is payload data). This provides us with HK monitoring every 2 days, plus 100% of our stored temperature measurements (taken every 1.5 hours). ThermoSat is a go!

More Realistic Bandwidth Allocation

A typical CubeSat radio contact using ham radio frequencies (such as Compass-1's 437.405 MHz) can provide from 1200 to 4800 bits per second (bps). To run the numbers, a 160-character text messages using 8-bit alphabetic data requires 1280 bits minimum (plus encoding and overhead), so you can send one to three texts per second using that higher rate.

To go into further detail, Compass One CubeSat (*http://bit.ly/1K6bkmV*) tells us, "The satellite receives commands in DTMF tones on the VHF uplink frequency of 145.980 MHz. Downlink of Ax.25 UI data packets is on UHF 437.405 MHz in either 1200 bps AFSK or 2400/4800 MSK, depending on the on-board setting."

Ignoring encoding, and still working with our simple analogy of *data via texting*, then, we can elaborate on our comm budget. Whereas before our rate was given *per month* as a total allocation, now we have a rate of *per second*. We need to now determine how many seconds we actually can communicate with the satellite.

Let's use a test scenario assuming one dedicated ground station. If we can communicate with the satellite during 5 of its 16 orbits, during which it is available to our single ground station, then we get 5 contact passes per day. If each contact pass lasts for 10 minutes (a reasonable estimate), we can now calculate both sides of our comm budget—what we have and what we need.

Per day, we now get 5 passes · 10 minutes per pass · 4 chirps per second (assuming the higher rate) = 12,000 chirps per day! If we can use all of that bandwidth, does this increase now enable both ThermoSat and PhotoSat to operate?

An upgraded ThermoSat comm budget is easy to calculate. We said we needed 45 chirps for all uplink and downlink to get data captured once per orbit. Now we have 10,000 times as much bandwidth (nearly exactly), so we can take 10,000 times as much data! Instead of one temperature reading once per 90-minute orbit, we can take a temperature reading at nearly twice per second. ThermoSat is now a high cadence LEO temperature probe.

More realistically, our detailed comm budget has to incorporate the balance between commanding, HK, and data. Instead of one-third commanding, one-third HK, and one-third payload data, though, we find more of our bandwidth can be dedicated to data. Let's run the numbers.

We are saying 5 passes per day, 30 days per month, so we have 150 passes. Each pass needs one command uplink to say "start transmitting," then one HK downlink. Then we'll get the data. We want to add some finer control and we have spare bandwidth, so let's add two more items:

1. The satellite will send a specific message once it is done transmitting all the data, to let the ground know it is done.
2. The ground station will uplink a *stop* command at/near the end of the contact pass to safely remind the satellite to stop transmitting and go back to normal data-collection mode. This stop command will include the number of *chirps* of data sent, so later on the ground we can count and make sure we got the exact number the satellite sent.

This adds only two chirps per pass, so it's not a large addition and it gives us more robust operations. We're not just telling the satellite *go*, we're also letting it tell us when it's done (so we are assured we got everything without missing any data) and we're informing it when we are about to no longer be able to communicate. These are our AOS and LOS markers. Let's look at our comm budget then per pass:

Available
: 10 minutes at 4800 bps aka "4 chirps per second for 600 seconds," or 2400 chirps per pass.

Needed
: 2 up (*start*/AOS and *stop*/LOS) and 2 down (HK data and *done*).

This means we still have 2396 *chirps* per pass. So, unlike our previous case, where we used one-third of our bandwidth for commanding and one-third just for HK data, we now have a more realistic scenario where the bulk of our comm time is going to be used for downlinking data.

For the ThermoSat example, we have bandwidth enough to downlink essentially 12,000 chirps of data per day, and since each chirp can contain about 40 data measurements, we have enough in our budget to downlink 480,000 temperature readings—enough to capture three temperature measurements per second. In fact, that's likely more than we need for our science goals. In essence, we have excess bandwidth. Even if we drop from 4800 bps to the slower 1200 bps rate, we still are capturing roughly one measurement per second with ThermoSat.

Now looking at PhotoSat, we are in the interesting state where a 512 x 512 color image takes about 22,500 chirps to transmit, and we are able to tweet just under 2,400 tweets per pass. So it still takes us over two days to transmit a color photo using our existing budget. We are still bandwidth-limited.

Encoding

In our previous examples, we converted or encoded our data (temperatures, images, HK information, commands) into *chirps*, then relied on our comm provider to send the chirps around for us. In satellite situations, you will have to encode your data to a known transmission format so that you can make use of the assigned communication frequency you've been allocated. Therefore, a little understanding of protocols (how things know how to talk to each other), encodings (converting data from one format to another), and packets (how the whole set of data gets chopped up to send it effectively) is useful.

The Internet works using a protocol called TCP/IP. When you send an email, send a tweet, request a web page, get an image file, all those get encoded and sent as TCP/IP data packets through the Internet connection. Whatever the content, there is a software (and often hardware) layer than handles the conversion from *real* data to TCP/IP and back. You don't need to know anything about TCP/IP to use the Internet. Your applications handle that for you (thankfully).

Satellite and ham radio, similarly, use protocols and encodings such as AX.25. You don't need to know much about AX.25 to use it. Instead, you need to have either your hardware (on the satellite) or hardware/software (on the ground) handle your encoding (converting your data to AX.25) and decoding (converting the signal back to your expected data).

Satellites have to be efficient and use minimal programming, so some knowledge of data encoding is useful in the design stage. Also, different protocols and encoding standards are used for different ground and communication systems. Just as you can't plug a USB into a Cat-5 Internet port, you have to make sure both ends of your system are expecting the same format.

So, like any good engineering decision, you should know enough to evaluate how you are slinging your data around. Reasons to choose a particular system include:

- It works for your particular communications setup.
- It's already built into your hardware and software.
- It has good support and documentation.
- It's easy to implement.
- You've worked with it before.
- It has the highest efficiency for your mission.
- It has the highest reliability for your mission.
- You can afford its cost.

You can evaluate this with a formal trade study that lets you rate how each solution fairs for each option above, or simply decide which factors above are most relevant and base your research on that. In the next session, we'll look at some formats used for other satellites.

Link Budget

In addition to working out our data usage with our comm budget, we need to calculate the amount of data our radio signal can carry, its capacity. This is the link budget for your comm system. This is—for a given broadcast frequency, transmitter strength, and choice of antenna—the bandwidth (rate of data) you can send over a given distance. This provides one of your fundamental limiting values and answers the question, "How much data can I transmit?"

By analogy, if you are talking to someone nearby, you can speak rapidly and they will understand each word. As they move further away, you either have to speak louder, or more slowly. Close up, you can rattle off a complex instruction in just five seconds like *you're about 50 meters from falling off a cliff so you may want to slow down* and they'll get it all. By the time they are very far away, you can at best communicate with single, blunt words like *stop ... cliff ... behind ... you.* How did sending 16 words at close range turn into only using 4 words at longer range? At greater distance, your link budget was smaller.

Units

The typical satellite unit has a bit rate in "bits per second" (bps). This is related to your *baud rate*. Baud rate is a measure of how many pulses (per second) or signals (per second) you can transmit. For sending digital signals such as we intend, a binary signal is a bit and is either zero or one. For digital signals, 1 baud = 1 bit per second, which makes things easier.

The base unit is bps. Higher data rates are in kilobits per second (kbps). If you are using a commercial provider, you might attain megabits per seconds (Mbps), and laser communication might someday provide gigabits per second (Gbps). For most amateur satellites, we are squarely in the bps and (maybe) kbps range. A typical ham data packet rate might be 9600 bps aka 9.6 kbps.

When would *baud* and *bps* be different? Using an encoding scheme will mean that a given signal takes more than just on/off bits. If your signal is not binary, but uses a variable signal using a continuous range of values—a modulation—to interpret, your baud rate will be higher than your bit rate. Sending a sine wave tone with pitch information, for example, breaks the *bit* concept since a single *pulse* also carries tone information.

For example, you can get more data per second sending a trinary signal. A positive pulse as +1, no pulse (0), and a negative pulse (-1) means you are sending more data per second. The use of such modulations allows for higher baud rates. Modulations include frequency (pitch), amplitude (volume), and phase (timing), all of which are a bit beyond the scope of this primer.

In practice, if you end up using modulation, you will implement a known modulating method that is:

1. Accepted by the radio-using community
2. Works for your given frequency range
3. Already implemented for you in hardware or software

You also will use a data coding (or encoding, I've seen both terms used) that ensures each bundle of data or *data packet* can be translated at the end. The reason you use known schemes is they are proven, they work, and they are consistent. Imagine one person is sending Morse code pulses (0-1) and the receiver is expecting piano tones (frequency modulation) and you can see the problems arising if unknown methods are used.

Back to bits. Since a bit is just one/off, to send useful information, you need to translate those bits into data units. Morse code uses patterns of multiple dots and dashes, for example (corresponding to on/off), to transmit letters. Most data is numbers, and you can represent numbers using bytes. A byte is 8 bits, and lets you send the integers in the range of 0 to 255 using just 8 bits. If you add more bits, you can expand your number range. You can cover the integer range from 0 to 2^N using N bits. Note that is N = 1 (1 bit), you're back at the range of 0-1!

Bytes is the size scale used for computer memory and hard drives—kilobytes (K, 1000 bytes = 1 K), megabytes (M or MB, 1000 K = 1 MB), gigabytes (1000 MB = 1 GB), and so on. To be a bit confusing, we are used to data storage as bytes (a 100 MB jpeg image file, for example), while transmission is one-eighth that because it uses bits (it takes 800 mbits to send that image file). *Bytes* uses capital *B* and bits uses either lowercase *b* or the word *bits* to distinguish the two.

Eventually, to determine the amount of data (total) you can transmit during a contact pass, you will multiply your data rate times the duration of your command window. A rate of 9600 bps for a 10-minute transmission window (600 seconds) = 9600 bps * 600 sec = 5760000 bits = 5760 kbits ... = 720 KB, or 0.72 MB of data, or the size of a typical low resolution picture (8 byte quality, RGB color, 480x480 pixels in size).

Your first takeaway should be that this is why you will not be streaming video if you are using amateur bands at 9600 bps. To avoid over-promising on the data you can send, you need to calculate your limits.

Link Budget Equation

The basic link budget equation is simply a measure of how much power is received, compared to the power transmitted:

Received power = Transmitted power + Gains – Losses

Units are decibels (dB), a logarithmic unit such that a difference of 3 dB is an increase of double the value. So adding +2 dB is not quite doubling the value. "dB" is the general physics equation, comparing two arbitrary signal strengths. For radio work, we'll also define "dBM," which is "dB of our signal versus a 1 milliwatt reference signal." Generic *dB* is "difference," and specifically *dBM* is *difference between our signal and 1 milliwatt*.

Key items involve the distance to your satellite, the choice of antenna, and the power levels being applied. You start with your raw transmission power, but that is just the starting part of the equation. As mentioned in Chapter 2, radio behaves under a distance-squared law. If you double the distance, the power received drops to one-quarter. Gain is a measure of how much boost an antenna provides in the chosen direction, and lets you minimize losses (at the cost of requiring more accurate pointing) at either the transmission and/or receiving antenna end. Finally, different frequencies have different efficiencies in delivering data—generally, the higher the frequency, the more data you can deliver for a given fixed power. All of these factor into the link budget.

The formal equation for dB is *dB = 10 log (Power / Reference level)*.

For most radio, we use 1 milliwatt (1 mW) as our reference level, making the equation for dBm: *dBm = 10 log (Power/1mW)*.

A watt is 1000 kW = 1,000,000 mW, or about 62 dBm. An amateur radio satellite transmitter, from space, is typically limited to 1 W or less, or about the power of a typical cell phone. In dBm, then, amateur space transmissions are typically going to be in the 500 mW to 1 W range, or 27 dBm to 30 dBm range.

For our link budget, we need to figure out our gains—due primarily to clever antenna design—and our losses—also due to our antenna, plus the distance the signal has to travel (propagation loss), plus absorption losses moving through the atmosphere, plus the distance-squared law for all signal drops. A common form is, given all losses and gains in dBm:

Power(received) = Power(transmitted) + Transmitting antenna gain + Receiving antenna gain - Transmitter losses - Receiver losses - Path propagation loss - Path absorption loss - miscellaneous losses

Antenna gain and loss is covered in Chapter 7, and is by analogy like the difference between a light bulb that shines in all directions (easy to see, but faint) versus a flashlight that uses a cone-shaped mirror to focus a directional beam (bright but only if you're facing it).

The free space path loss, simplified, looks at how a radio signal moving outward is like the ripple from a stone dropped in water—the waves spread out in a sphere and therefore the total power spreads out over a larger area, resulting in less power per square centimeter as it moves further out:

FreeSpacePathLoss = (4 d/w)^2

In dB, the free space path loss (FSPL) can be split out into the distance, frequency (fw = c, with *f* in Hz), and that *4 pi* part with:

(FSPL (in dB) = 20 log(d in meters) + 20 log(f in Hz) - 147.55)

or, in more convenient units:

(FSPL (in dB) = 20 log(d in km) + 20 log(f in MHz) + 32.45)

where d = distance between transmitter and receiver, and w = the transmission wavelength. Immediately this shows that different radio frequencies have different propagation losses. Since your satellite is moving, your final link budget will be in chart form, indicating miminum and maximum budgets for each part of the period during which you communicate with your satellite.

Going Deeper

At some point, you will discover there are online link budget calculators that spit out the exact numbers you're looking for. It is useful to know what data to input, so hopefully this book bridges the gap between newcomer and expert.

The key to a link budget is the concept of the signal-to-noise ratio (SNR). If there is more noise than signal, you can't tell what is going on. If the signal is above the noise, you can hear it. The less noise, the more clear the signal. You typically want a S/N ratio above 3. SNR is literally the ratio of (signal/noise), but is often given in decibels as *SNR(dB) = 10 log(signal/noise)*. The higher your SNR, the more data you can transmit across your system.

A strong cellullar *3-4 bars out of 5* signal typically has an SNR above 25 dB (*http://bit.ly/1E5rch8*). For any application, you want the highest SNR you can get. For nanosatellites, you are usually strongly constrained in terms of wavelength (license-limited) and transmitter power (power-limited and license-limited), so the only way to boost your SNR is at your ground station. This is because you have more design choices and fewer power and weight restrictions in building your ground station. Typically you can improve SNR through effective, efficient antenna choice—a large area directional ground antenna pointing accurately at your spacecraft is the largest area for improving your SNR.

We'll add in a term, the power of our antenna. If you have a 30 watt source, that means the power is 30 watts. This could be anything putting out energy per second, since watts is just "energy per second." An engine, a light bulb, a radio, they all put out energy, so they all have a measure of power. Physics is handy that way, it doesn't care about what *flavor* the power is, it's just energy per second. For radio work, we specify power in watts but also talk about the shape that power radiates as. Just like a light bulb casts different illumination when you compare a bare light bulb shining in all directions to a focused light bulb in a flashlight, so radio power depends on the antenna *focusing* and shape. We'll start with the basic raw power, though, the effective isotropic radio power (EIRP), or power assuming it is radiated evenly in all directions.

In general, *Power(actual) = Power(out) - Losses*. You just start with your basic power (EIRP) then subtract all loses. Losses for radio power are many and varied, the primary ones including the aforementioned free space path loss (FSPL) that includes both the loss due to distance, and a frequency-dependent loss component, added together (in dB), as given before. Also, there is noise in your receiver antenna—often characterized as a *noise temperature* T—and the miscellaneous *other losses* that we'll briefly cover in a moment.

We also have to factor in the gain or increase in your system due to effective shape or focusing. A flashlight, for example, adds a *gain* to the light power in the direction of the beam, at the cost of diminishing the light of everyone outside of that path. We use gain in systems to provide stronger signals, in essence amplification.

Taking all of these together, your SNR is given as:

(SNR = EIRP + G/T - FSPL - Loss(other) - k - Data rate)

where G is the receiver antenna gain, T is the system noise *temperature* of the receiver (basically, receiver noise), and k is Boltzmann's constant (as –228.6 dBW/KHz).

If you calculate your receiver SNR—independent of what is transmitting or what losses occurred in transmission, just looking at the receiver—as SNR(Receiver), you can compute your data rate by rearranging the equation:

Data rate = EIRP + G/T - FSPL - Loss(other) - k - SNR(receiver)

Given that EIRP is set by your nanosatellite radio power and nanosatellite antenna design, that is relatively fixed. The free space path loss is set in stone by the physics of what wavelength you are using and how far the satellite is. "Loss(other)" has many factors that are mostly out of your control, such as rain, dust, electronics implementation, atmospheric absorption, and current solar activity. *k* is a fundamental constant and unchangeable. The SNR of your receiver is likely already minimized in terms of components purchased and electronics design. The antenna noise *T* is—assuming good implementation of the hardware—relatively fixed, just like receiver SNR.

This leaves the fundamental *tweakable* parameter as *G*, the antenna gain. A single dipole has a typical gain of 1.66. A Yagi antenna is basically a stick with multiple dipoles attached, and has a gain of roughly 1.66 * the number of elements... as long as each dipole on the Yagi is separated so they are in phase (receive and respond to the signal at the same time) and thus separate by one-half, one-fourth, and one-eighth the wavelength being received. A typical Yagi antenna will have a gain of 10dB or so. A dish antenna, for comparison, has a gain proportional to its area. If you double the area, you double its gain. As you can see, antenna shape, size, and design are a very important part in ground station design.

Regular dipole omnidirectional antenna that radiates signal nearly evenly in all directions (except *up* and *down*), is shown here:

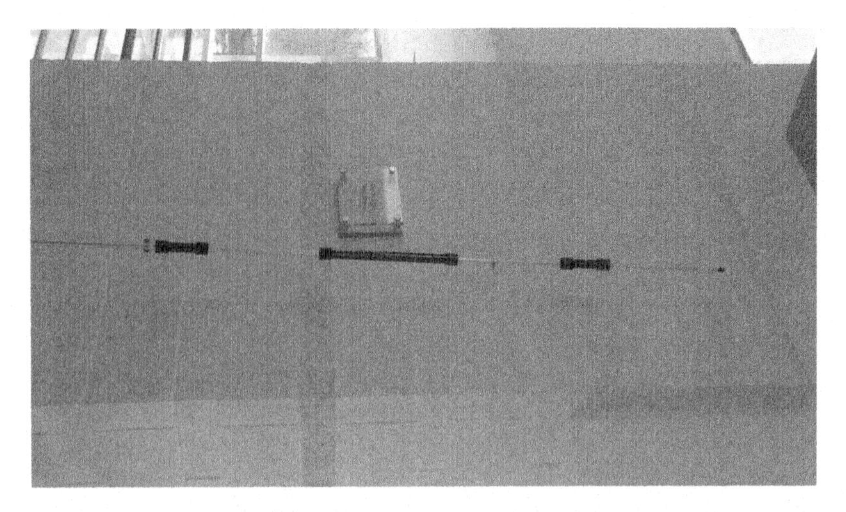

...compared to a focusing dish-type antenna, that has higher gain in the direction it faces but lower gain in all other directions:

It is key to understand that your link budget is not a constant, but varies in time as your satellite moves. You generally have a minimum value at the AOS, when the satellite is at its geometrically furthest difference (long free path length, more atmospheric obscruction). It increases until the satellite is nearly overhead, then drops as the satellite sinks towards the horizon and heads for LOS. Each orbital pass will have a slightly different link budget, because each pass has a different altitude and angle relative to your station. Add in weather effects, and you can easily expect a 10 dB or 20 dB shift in your link budget. Because the link budget determines your maximum data throughput, you should calculate an average link budget as well as likely minimum and maximum link budgets. Expecting imperfect comms and assessing likely ranges for throughput are needed to ensure you aren't trying to send more data than your communications channel will deliver.

Data Budget

Use your orbital elements to predict your number of contact passes and the duration of each contact pass. Multiply that time by your link budget rate, divide or multiply by your encoding scheme, then compare this total data amount to the data you wish to transmit. Book 3, *DIY Instruments for Amateur Space*, goes into depth on how to calculate data volumes you want to bring to ground.

If you want a quick estimate for an amateur band LEO satellite, assume you can get 9600 bits per second over a 5-minute period for five consecutive 90-minute orbits, followed by 10 orbits with no contact. That's a fairly standard LEO profile for a single ground station. At 8 bits per byte, assuming 8-bit data numbers, that's 1.8 MB of data transfer per day. You can split that between uplink and downlink as you wish.

To get effective data down from a satellite, then, you need either multiple amateur ground stations, access to a higher rate commercial comm system, or access to an always-visible communications relay constellation. Amateurs are currently working all three approaches.

Alliances of amateur ground stations have been forming for individual missions. Once you've designed your system, you can network with other amateurs to have them support your satellites. Many hams with existing gear are happy to listen in and even send you any data they fetch if you provide them with the *keys*, the necessary frequencies and access codes to get the data. Systematic networks that anyone can tap into via Internet are still in testing, including GENSO, CarpComm, and others. By the time you read this book, hopefully someone will have solved the *build it and they will come* universal access CubeSat ground network for amateurs.

The communications constellation approach relies on bouncing your signal from ground to a network of satellites, hopping the signal to your satellite. You ship your uplink data via Internet to the provider, they transmit through their satellite network to your satellite, and then they rely your downlink back through their network and deliver your data to you via Internet. A TDRSS or Iridium relay can give you 24 hours of potential coverage, although you'll pay per minute for usage. Typically, they use higher power transmitters and our initial calculations show you ironically end up with low data rates because of your power budget, not your comm budget. Put simply, even a 3U CubeSat charging via solar can only power an off-the-shelf Iridium transponder for perhaps 20 minutes per day. However, as more commercial entities get into the nanosatellite communications market, I expect comm unit power designs to move to lower power, access to networks to be more easily available, and effective data rates to therefore greatly increase.

7/Antenna Design

Radio waves are actually photons. We almost always deal with their wave-like properties, but we can discuss them the same way we discuss light. First is that, if you focus a light source, you increase its power in the direction focused, at the cost of reducing its visibility from other sides. This is a key point of antennas. Second is that, like camera lenses and telescopes, larger antennas tend to capture more radio signal than smaller antennas. Finally, as with audio speakers, you need to make the fundamental size of your antenna match the frequency of the radio waves you are using.

Fortunately, your task is easier than that of a general radio user, since your satellite will be broadcasting at or near a single specific frequency. Therefore, once you build and match your transmitter and receiver, you do not have to worry about varying it later. Although you have to do some initial hardware research, once you build your setup, it will be stable.

Antenna Shape

We call a standard rod-like antenna a *dipole*. There are also circular antennas and other designs, but the bulk of satellite work uses dipoles. An antenna is a dipole because you apply voltage at one end so it has a positive and negative point.

Magnetic dipoles

Magnets are also dipoles because they have a plus at one end and a minus at the other. Dipoles abound in physics and in nature.

An antenna can be *omnidirectional*, which means that it broadcasts and receives signals from all directions at low efficiency. Or the antenna can be *directional*, which means it boosts its efficiency in the direction it is pointing, but has really poor performance in any other direction. A dipole antenna is a *nearly omnidirectional* antenna, tending to send signals strongly all along the horizontal plane, but sending not as much signal in the up-down direction.

In satellites, these are often called low-gain and high-gain antennas. The omnidirectional low-gain antenna doesn't require any specific pointing, but is weaker. The high-gain antenna only works if pointing in the direction of your ground station, but when it works it has a much stronger signal. The primary difference is shape. A high-gain antenna involves shaping the antenna to focus radio signals in one direction (gain) at the cost of having less radio in all other directions.

This kind of big high-gain antenna, the Capitol Technology University *Big Dish*, is what we want:

whereas this handheld rig is more like what we can get into space: a small, low-gain antenna:

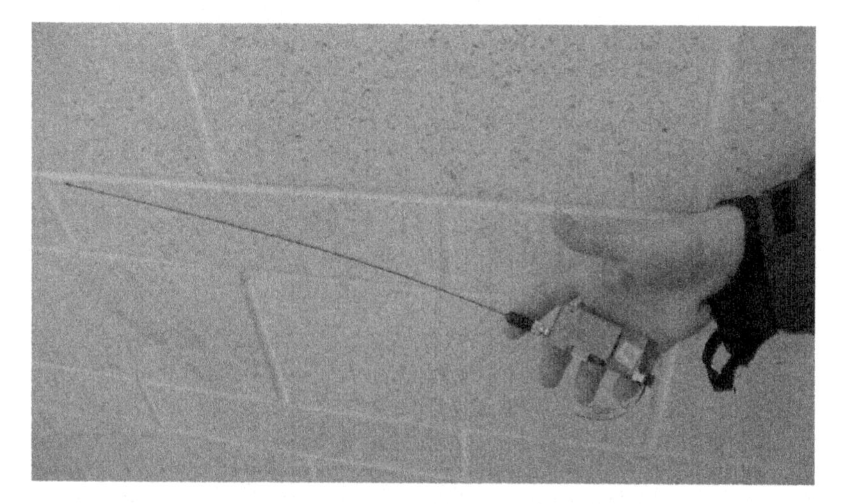

An omnidirectional antenna or low-gain antenna or dipole antenna can be just a single metal bar or strip. A car radio aerial or the antenna on a walkie-talkie are examples of (nearly) omnidirectional antennas. In practice, a straight aerial of that sort isn't fully omnidirectional. Instead, it captures signals coming from the full 360 degrees around it, but is a poor receiver along the direction it is pointing.

This leads to the basic rule of directionality. An antenna works best if roughly *broadside* to the signal. If you point a straight rod straight at your radio *target*, the signal is weak because you don't have a large capture area for the radio waves. If you put the antenna so its entire length can receive radio waves from the direction of the source, you are maximizing your signal. In this way, antennas behave more like sails on a sailboat—the more your antenna has a *capture area* facing the direction of the signals, the more signals your antenna absorbs. A dish or a Yagi are examples of high-gain antennas.

Common Spacecraft Antenna

It is most likely your satellite will use one or more low-gain omnidirectional antennas to broadcast uniformly in all directions, while your ground station will use a directional antenna that actively tracks and points at the satellite. This makes the satellite easy to use—you do not have to worry about what direction it is facing—while improving response at the ground station—which can afford the cost and weight of fancier equipment.

Most amateur satellites do not have active attitude control, which means they lack the ability to control which direction they face. Therefore, most amateur satellites use one or more omnidirectional low-gain antennas to broadcast in all directions, accepting that the signal is weaker but ensuring that the satellite can be heard from the ground regardless of how it is facing.

Even a single antenna will *see* the ground often. A common configuration is to stick two or four antennas off the *base* of the satellite, so that even if the satellite is tumbling, at least one or more antenna will have a good facing to the ground. In many ways, having consistent reception is more important for CubeSats than having a high efficiency in reception. It is better that the data always comes through, even if slowly, than to have dropouts that lead to gaps in the data.

Most NASA and commercial missions, in contrast, have attitude control and specifically use a high-gain antenna that is always pointed towards the ground station; to ensure stronger radio signal power—which translates to higher data rates. As a backup, they all include a low-gain omnidirectional antenna, useful if there is a satellite problem that errs and points the high-gain antenna the wrong way. Low gain = reliable, high gain = efficient.

Common Ground Station Antenna

Meanwhile, on the ground, for satellites, you will likely skip the omni and instead use a directional antenna such as a Yagi. A Yagi is one common design for a directional antenna, and looks a bit like the love child between a weather vane and an arrow. You point the antenna support in the direction you need so that all of the *vanes* that are sticking out sideways can draw signal. It gets great performance, but signals to the side are essentially ignored. A handheld Yagi antenna from the side:

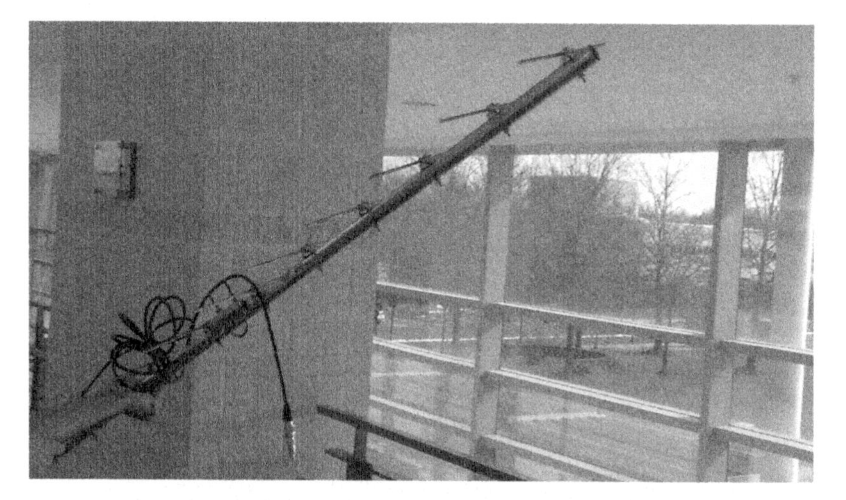

and rotated to show the top:

In both cases, the strongest reception is in the direction the arrow-like antenna is pointed, and it gets very poor performance from the sides.

If your ground antenna is directional, you need to know where your satellite is so you can point (orient) your antenna properly. This involves three steps:

- Directional antenna mounted on a motorized, computer-controlled mount
- Software to control the antenna rotator
- Orbital elements for your satellite so you can deduce where to point

You can often do these items by hand. If you know, for example, that the International Space Station (ISS) is going to rise in the west at 9:10 and set in the east at 9:14, you can easily manually track it. There are many websites that list how to visually spot the ISS, so I recommend you first do that before trying to track it with radio. The technique is the same:

1. Look up the transit position for ISS.
2. When it appears (location and time), wait until it comes over the horizon, do the initial Acquisition.
3. Track its eye as it moves.
4. When it falls below the horizon or moves into the Earth's shadow, you are done.

Two sites with more ISS (and other satellite) tracking information are Spot the Station (*http://spotthestation.nasa.gov*) and SATFLARE (*http://satflare.com*), among others. Satellite spotting is a fun activity and the Internet is full of helpful predictions to tell you when and where to find all sorts of satellite activity.

If you're interested, also look up *Iridium flares* for details on spotting the tiny yet shiny constellation of Iridium satellites.

Once you've done the spotting visually, you are ready to try radio spotting. Unlike visual spotting, which can be done only at night or twilight, radio spotting can be done day or night. Armed with your direction antenna, you will:

1. Look up the transit position for ISS.
2. When it appears, point your antenna at the horizon and wait for AOS.
3. Track ISS so that you always have a strong signal.
4. When ISS falls below the horizon (LOS), you are done.

The only difference between the two methods is that you aren't visually spotting the space station, but using the radio signal strength to keep on target. This is easier to do than describe. You just keep moving the antenna and if the signal gets weaker, you are either leading it by too much, or lagging too slowly. Despite the name *directional*, a decent Yagi has a broad enough *beam* that you do not need high accuracy to pull this off.

The first time you track ISS, try doing it at night so you can also visually spot the station. Once you get the hang of it, it's easy to do anytime. If you can track the ISS, you can track a satellite. Once you build the fancier automated tracking rig mentioned above, you can test it by robo-tracking the ISS with your new setup.

Antenna Size and Frequency

The second quality of an antenna is that it has to be built to the frequency (or wavelength) of radio it can pick up. If you hit the different blocks of an xylophone, each makes a distinct tone or pitch—the longer blocks make lower pitches, the higher blocks make higher pitches. Antennas similarly have to be the right size to resonate and pick up their specific radio frequencies.

The rule of thumb is your antenna length should be a ratio of the wavelength you want to receive, ideally half the wavelength, if not some power-of-two fraction of it. The reason is you want the incoming radio wave to *drive* the antenna at its ideal frequency, so it can deposit the most power into it.

Impedance and SWR

The material and the cables connecting the antenna form a circuit. All circuits have some amount of resistance and some impedance. Resistance is exactly as it sounds, resistance to the flow of electricity. Even the conductors we are using have some resistance (superconducting antennae are left as an exercise for the reader). Impedance is, roughly, the sluggishness in how quickly the antenna or circuit responds to a change in voltage or current.

An antenna is made of a material (metal) that reacts strongly to radio waves, cut to the length so that the wave maximizes its driving influence. Without going too deep into antenna theory, there are nonresonant and multifrequency designs as well, but for CubeSat low-power operations, we want to make an optimized yet simple antenna that maximizes signal and minimizes losses.

The term *impedance* is used as a measure of how closely changes in voltage are matched by the change in current. If you apply a current (signal) to an antenna and the voltage changes in phase, the antenna power is being radiated (or, if a receiver, absorbed). If the response of the antenna lags the voltage change, the power of the signal is stored in the antenna's electromagnetic field instead of being radiated out. The ideal impedance is neither too large nor too small, and is related using a term called standing wave ratio (SWR).

SWR is a measure of your antenna's efficiency, and is the difference between the transmitted *forward* energy and any stored or *reflected* energy. We also introduce the term isotropic effective radio power (ERP), which ideally will be 100%, that is 100% of the energy you are pushing into the antenna manifests as radio waves going in all directions. (For now, we are ignoring antenna shapes and directionality.)

SWR = Forward - Reflected + (Reflected/Forward)

SWR is measured with (of course) an SWR meter as part of your radio gear, usually as a separate box installed between the antenna and the radio. You will use an SWR meter for your ground station. You will temporarily hook one up to your satellite while testing and tuning your system, but definitely remove it before launch. There is no point in flying a meter in space that no one can read!

SWR is given as a ratio (e.g., 1.0:1 = 0% loss, 100% ERP). A SWR of 1.0 indicates the antenna is perfectly matched to your signal line. At 3.0:1, there is 25% loss so the ERP is 75%. At 6.0:1, you are at 51% loss and 49.0% power, and higher than that isn't worth using. Most advice says to try for 1.2:1 or less, and anything above 3 is usually marked in red on your meter.

(What happens to the nonradiated *stored* power? It tends to cause interference in both the connected electronics and via radio emission at other frequencies dumping into nearby other electronics. Not good.)

The hardware to match impedance is technical and varied, including antenna tuners (a powered box to help match impedance), choosing the right wire, balun transformers, and proper choice of wire. At this point you should look at the *ARRL Antenna Book* or any of the fine web ham resources for exact design. Antenna design depends on frequency, use, and hardware limits.

Antenna Materials

ERP_ = (Power of transmitter) - Losses + Antenna Gain

versus a dipole antenna.

EIRP = (Power of transmitter) - Losses + Antenna Gain

versus an ideal isotropic antenna.

Many amateur CubeSats use an onboard antenna consisting of two or four wire strips, cut to the length of one-half, one-fourth, or one-eighth the wavelength (depending on their frequency). The *70 cm band* is a common range, typically using frequencies of 430-450 MHz (depending on each country's licensing). A one-half wavelength antenna is therefore 35 cm, or just under 14 inches.

These straight antenna can be made from a hardware store metal tape measure, the roll of metal ruler stored in a box. One plus of using a tape measure is you don't have to use a separate ruler to cut off just the right length—but you should remove any paint or coatings from the metal. An antenna works best if it is the bare metal, with no covering, like the tape in this tape measure (photo credit: Evan-Amos):

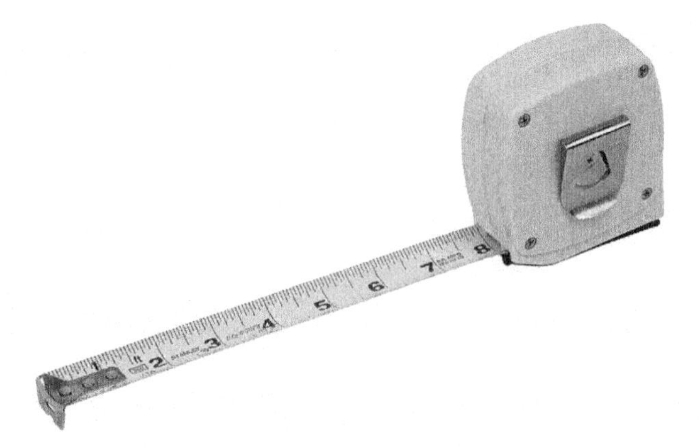

A bonus of using tape measure antenna is that you can (and will) roll the antenna up into the satellite for launch, and rely on the fact that it will spring back out to its full length once the satellite is kicked away from the launch vehicle. Presto, antenna and deployment system already designed for you.

Using that design, you can easily deploy four antenna in a cross pattern, or two antenna as opposite sides. For material, you take a cleaned, unpainted, stripped metal tape measure, cut off the proper length, then roll it up inside the cube before launch. When the satellite launches, the antenna immediately uncoils to its straight length, then sits there happily until you eventually get around to powering up the satellite.

Fancier antenna exist, including coiled designs and models made specifically for CubeSats. The principles are the same: match the antenna to the wavelength (in size) and the circuit (based on the frequency and the hardware and connecting wires you are using). Build a ground station that can receive that same frequency. Allow some frequency variation for Doppler shift. Test, test, test. Then launch.

A short summary for a typical amateur satellite is:

- Omni for satellite: tape measures
- Directional for ground: Yagi on a motorized mount, computer, software
- Orbital elements: from web

Power, ERP, and EIRP

We talk about the *power* of the antenna, for example, that we are using a 0.5 watt transmitter, so that's our transmitter power or output power. Effective isotropic radio power (EIRP) and effective radio power (ERP), mentioned earlier, include any losses due to the circuits and wiring, and any boost if you are using a directional antenna.

The ERP is calculated, in dB, as the power minus the losses (as usual) plus gain, and is compared for your antenna shape against a hypothetical *dipole antenna*. It's a measure of your actual power, versus the lower power you would have if you just used a dipole antenna—specifically, an ideal half-wave dipole antenna that provides a little gain. Because we are using dB, a logarithmic unit, we're *adding* each component:

ERP = (Power of transmitter) – Losses + Antenna gain

versus a dipole antenna.

The EIRP is nearly identical, but tells you your power relative to an ideal isotropic antenna—which has no gain or signal boost in any direction—instead of a dipole (which has a little gain in some directions):

EIRP = (Power of transmitter) – Losses + Antenna gain

versus an ideal isotropic antenna.

Both of these are ways of saying how your specific antenna shape—which has a gain—compares to a low-gain (dipole) or isotropic (no gain and radiates equally in all direction) antenna. So both are slightly larger than those lower-efficiency designs. Note that if you had a theoretically lossless circuit and an antenna with no losses that radiated its power perfectly isotropically (in all directions), the ERP would be equal to the transmitter power. You can think of transmitter power as *power applied* and ERP or EIRP as *maximum power delivered*.

From a licensing and use level, this means both ERP and EIRP are the power a receiver can, at best, expect to face. It is therefore a very useful measure. While transmitter power says what the circuits put out, ERP says what the circuits + antenna actually are able to deliver. It's a more useful number when you're trying to figure out what to expect at the receiver.

So to use the Wikipedia.org "Effective Radiate Power" figures as an example, if your transmitter is putting out 0.5 watts, and your circuit + antenna have a 6 dB loss but the directional antenna provides a 9 dB gain (for listeners facing it), then the ERP = +9 dB – 6Db = +3 dB ERP gain, or roughly a 2x boost on the 0.5 watts, so the ERP = 1 Watt. Notice that, if you're not facing the antenna, you are receiving far less than this, but ERP is usually assumed to be the maximum power anyone might experience.

As a side note, ERP = EIRP/1.64, because 1.64 is the gain difference between an ideal dipole versus an ideal isotropic antenna. Put another way, an isotropic antenna has no gain (amplification factor = 1), while a dipole has a gain of 1.64 (amplification of 1.64) if you face it just the right way. From a functional level, ERP is important for licensing because if you exceed the ERP allowed, you are violating the FCC (or your local country's equivalent) rules and have to shut down your transmitter.

Existing Antenna Networks

We covered the TDRSS, DSN, and NEN earlier, and this chapter discussed single antennas. A short list of different existing antenna systems can be researched further by looking up the following:

- Geosync: fixed ground antenna
- Tracking and Data Relay Satellite System (TDRSS)
- INTELSAT, INMARSAT, Galaxy, DSCS, FLTSAT
- Deep Space Network (DSN), Near Earth Network (NEN)
- Single antenna

8/Performance Characteristics

Data, Error Rates, and Availability

At a high level, your ground-to-space and space-to-ground system has five factors: data rate, bit error rate, end-to-end delay, link availability, and security. Security includes antijamming, antispoofing, verification of data integrity, and potentially (but unlikely for CubeSats) encryption.

The data rate is discussed in detail in pretty much every other chapter. It includes the issues of signal strength, bandwidth, data packet format, and many other considerations. It is basically the end goal of your comm system.

The bit error rate (BER) is the probability that some data will have errors. A BER of 10^3 means that 1 in 1000 bits will be incorrect. Voice communications systems can usually handle that level of glitchiness. Compressed data, which often uses algorithms that require 100% data fidelity, require BERs of 10^{10} or better. For most CubeSats, where data losses are acceptable and compression and encryption are rarely used, we can go with a BER of 10^3.

But there's more to data fidelity than just the bit error rate. How can you can tell if data packets are bad? Can you request a retransmission of data if you determine some of it was bad? BER and data fidelity is even more crucial on the uplink of commands to the spacecraft and in the downlink of status messages on the satellite health, than in the downlink of spacecraft data. One key quality of data is that not all data requires the same fidelity.

One way that BER can be accommodated is by choosing a data packet format that includes a periodic checksum. The checksum is a final number that is computed based on the values of the preceding data. The checksum is computed by the sender before transmitting the data, then included at the end of that batch of data. The receiver then computers a checksum based on data received, and compares that with the suggested checksum. If it doesn't match, there was a transmission error and a retransmission can be requested.

A very simple checksum might be, for example, that I send a sequence of four numbers plus a checksum that is "the tally of numbers that were odd." So the data set of *4 1 3 6* would have a checksum of *2* (two numbers were odd). I then transmit *4 1 3 6 2*—but assume there was an error and the receiver only got *4 1 3 5 2*. When they calculate their checksum, they get *3*, which does not match the transmitted *2*, so they know an error occurred, even though they do not know what the error was. It it sufficient to just know it occurred and therefore not act on the data sent, just flag it as "bad data" and ask for a retransmit.

In practice, checksums are more detailed but follow the schema given as an example. A checksum typically indicates whether a given block of data was successfully transmitted; the recipient can then decide how to handle any blocks that fail the checksum check. A typical response for data is *request a retransmit*; a required response for actual command sequences is *ignore and request a retransmit*. Again, different data has different fidelity needs.

Given that command uploads and telemetry downloads are usually much smaller in data volume than actual instrument or image data, we can still maintain a reliable system with our lower BER.

End-to-end (E/E) delay is almost irrelevant for most CubeSat missions. The E/E delay is the time between message transmission and receipt by the ground. Because CubeSats rarely do real-time commanding or live relaying of communications, delays of seconds are fine. In low earth orbit, there is minimal light speed lag—even in geosynchronous orbits 35,785 km up, the delay is only 0.12 seconds. However, there may also be processing delays or the need for a remote ground station to relay its data to a central command office. Designing a mission so it does not need any synchronous communication is therefore ideal to avoid requiring higher E/E standards.

If you are using a data packet protocol that includes error checking and automatic requests for retransmission of bad packets, the E/E rate will hinder the ability to get a timely retransmit. Retransmissions might have to wait until the next contact pass. One common method is to schedule your contact passes so the first items are both the smallest yet most important for mission health, closing out with the longer transmission of the larger science data.

Link availability is exactly what it sounds like: how often the ground and satellite can communication with each other. Link availability includes both orbital and schedule issues as well as unexpected interruptions. Schedule issues typical with LEO CubeSats is that a single ground station will have only a 4 to 10 minute window per orbit, at best, to talk with the satellite. Unexpected issues can include items such as thunderstorms that interfere with the radio signal, or power outages that knock the ground station out. For a tumbling CubeSat, link availability may also include whether the changing satellite orientation affects the radio signal strength due to antenna pointing, causing dropouts or weaker signals.

Security

Security includes communications security (COMSEC) and transmission security (TRANSEC). COMSEC protects the data, through encryption and checksums and validation. TRANSEC is ensuring the signals, once sent, arrive safely and intact between ground and satellite.

Encryption is usable if your frequency allocation or comm provider allows it. Amateur ham radio is not allowed to use encryption, on the basis that you are sharing a public resource. Licensed or private communications each have their own Terms of Service (ToS) on whether encryption is allowed. Bear in mind that encryption generally adds to your data overhead—makes your data bigger. If you encrypt, you will get less data down than if you sent it in the clear.

For TRANSEC, anti-jamming is ensuring that someone else cannot block your transmission, for example by broadcasting at the same frequency as you but with higher power. For CubeSats, this is not a concern we can address; our transmissions tend to be on shared spectrum and in the clear. We typically rely on national agencies such as the FCC and IARU to police any cases where someone accidentally or deliberately blocks our attempts to communicate with our satellites.

The TRANSEC issue of anti-spoofing and the COMSEC issue of verification of data integrity are more crucial issues, though currently both are rarely seen with amateur satellites. Spoofing is where someone either pretends to be a legitimate ground station to try to take control of your satellite without permission, or fakes being your satellite to send you fake telemetry and data.

Users will probably try to see if they can *hack* your satellite and operate it. Two ways around this are *security by obscurity* and by the use of validation codes (passwords). Security by obscurity means you simply don't tell anyone what your satellite commands are, so they can't send their own commands.

One way to implement this is to have predefined command mnemonics that trigger satellite operations. If you don't send a proper mnemonic, the satellite will ignore the command. So perhaps the sequence *chsx1001* is the eight-character mnemonic for *transmit health and safety now*, and *cmode1x1* tells the satellite to switch instrument 1 to a predefined *mode1*.

The downside of security by obscurity is that, if someone gets your list of allowed commands, they can now operate your satellite. The upside is that you will probably define mnemonics anyway, because they are a very efficient way to command your satellite. From a security point of view, though, they are not enough.

Passwords and validation codes are a second layer for security. A password is just like for a computer account—if the password isn't at the start of your transmission, the satellite ignores it. The downside of passwords is, if you are transmitting them *in the clear* (unencrypted, in plain text), anyone that finds them out, again, has access to your satellite. *Finding out* can mean listening in to your radio transmissions, or asking someone on your team (social engineering).

One security method is to have an algorithmic component to the password, similar to a checksum, such that even someone intercepting a message still will not gain access. A simple example might be to start each message with a two-number code, which is based on the time at the start of the communication (using the clock onboard the spacecraft). So if the spacecraft thinks it is 05:47 when the transmission is received, it would expect the code *54* for a valid ground command, otherwise it will ignore it. Of course, this limits you to only a 10-minute window to transmit that command.

Security is not my field, and there are much more robust schemes out there, but this hopefully gives you a primer on basic communications concerns. One final concern is that you should be sure to physically secure your ground station. Restricting physical access to documentation and protocols goes a long way towards maintaining communications security.

Eagle-2 Legit Hacking

The Pocketqube-format Eagle-2 picosatellite provided an intentional challenge for hams and AMSAT users. The team deliberately told people to *try to figure out the cool ways to command our satellite that we put in, but didn't document for you, then let us know you did it and win a prize!* The satellite was an engineering test project. Downlink by default communicated its call sign and/or a preset message using slow FM Morse code. After this slow Morse "beacon" is sent, there is a 10-second window to accept command uplinks—from anybody. The builders provided three legitimate commands that any ham user could use to get a report from the satellite. They also coded but did not disclose additional commands that can change the exact message being sent, and left discovering those as a radio challenge to enthusiastic hams. As of their 2014 presentation, the message has been changed three times.

Historical Satellite Hacking

While many commercial satellites use asymmetric public key cryptography as encryption, most NASA missions (because of their public mandate) send their data in the clear, with no encryption. This also saves on overhead. Commanding the sat is locked down, however, to prevent others from gaining access (hacking) NASA satellites. Nevertheless, NASA satellites *have* been hacked. Two notable cases are the Terra and Landsat-7 missions. They were hacked by the Chinese government, or agents acting on behalf of China. This was reported in 2011. The hacks were:

2007, Terra
: Disrupted operations for two hours and nine minutes.

2007 and 2008, Landsat-7
: Hackers gained enough access to command the satellite, but didn't.

For both satellites, the hackers gained access via the Spitsbergen, Norway commercially operated ground station.

A NASA computer system administrator once told me that NASA is a very popular target for hackers, because you get all the *street cred* of hacking a major government agency, but without the lethal armed response that most other agencies (especially Department of Defense) can bring to bear. So please support your local scientist, and don't hack NASA for sport!

9/Concept of Operations

ConOps

ConOps is your concept of operations, occasionally called Ops-Con. Mission Operations exists to perform normal operations and resolve anomalies. Operations has four core functions: call them Commanding, Monitoring, Trending, and Science. The Mission Operations Center (MOC) handles commanding the spacecraft and assessing the status of the spacecraft via HK data, as well as trending. The capture and archiving of the science data is the responsibility of the Science Operations Center, or SOC. The MOC and SOC can be the same entity, it's just that one deals with the satellite bus, the other with the instruments and data. From the hardware view, your satellite consists of a bus + payload. These deliver three streams of data, shown here with the uplink and downlink elements:

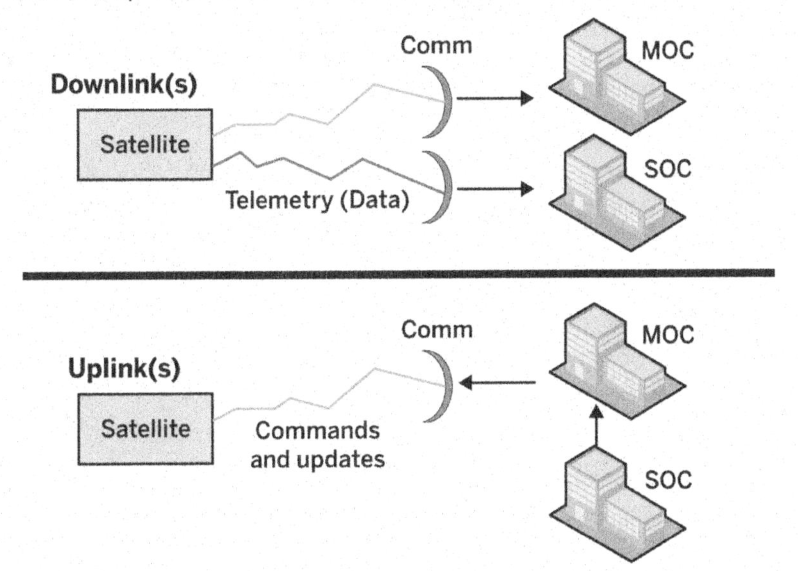

Uplink can be via one or two antenna ground networks. Housekeeping (HK) data goes to a Mission Operations Center (MOC), while science data goes to a Science Operations Center (SOC). HK telemetry is usually frequent and small; science telemetry is usually of larger size and therefore often less frequent.

The comm system may be one antenna or two (or more), relaying to the MOC and SOC appropriately and handling the uplink. Health and safety and science data may also use the same or different frequencies, depending on your ConOps.

There are many different architectures, ranging from *one ground station, one frequency, one satellite* to *multiple ground stations, in-orbit satellite relays, multiple frequencies, one satellite*.

For commanding, the SOC usually relays their requests to the MOC. For small missions like ours, the SOC and MOC may be the same entity—or even the same person.

In addition to communications, your ConOps should indicate who is telling you where your satellite is. Make sure someone is providing ranging and flight determination information and sending it to your MOC.

Good operations involves a team, each tracking a different aspect of the mission. Even if your team is one person, you are generally looking at only one aspect of the information flow—health of the satellite, quality of data, whether commands are being sent—at a time.

Some typical roles are the mission operations manager planner; scheduler; comm lead; bus monitor(s); and science monitor. Here's one set of typical roles:

Mission Operations Manager (or MOM)
: The person leading operations. They may just boss other people, or they may also handle a role below, but everyone needs to agree who is in charge.

Scheduler
: You need someone in charge of making overall mission schedules and detailed activity schedules, of scheduling data link times, and of overall planning. The three short-term

components they track are when the next data downlink will occur; what the next set of planned commands will occur; and when the next command uplink will occur.

Comms
Getting the data, processing it, and ensuring it goes to the right people is the next seat. Typically Comms will be making sure the telemetry is flowing, and also in charge of hitting the button to send any commands at the proper time.

Bus Monitor
This role analyzes the HK data and assesses the status of the satellite and its subsystems to ensure everything is working with designed tolerances. They assess whether the mission is alive or in trouble.

Science Monitor
This role checks that the quick look or actual data is valid and what is desired. They assess whether the mission is succeeding in its goals.

Navigation
If your satellite has active position control, either in attitude (facing) or orbit (maneuvers), someone in Ops should be in charge of tracking this.

SysAdmin
Maintaining your hardware and software is the final required role. They need not be in Ops all the time, but be sure someone is on top of your actual ground station hardware and software to ensure you are up and running when you need to be.

While any of these seats/roles can be filled by one or more people, it is important to realize the breadth of tasks required for operations. There is no one *standard* way to set up operations. Your team may just be you or a small group of friends, but understanding roles and responsibilities will lead to better operations.

XTE Staffing Example

The XTE (X-Ray Timing Explorer) had the following roles:

- Project scientist: end user, but also part of the staff and provides support from outside the mission
- Dev team shared by missions, both support and in-house staff
- Multimission Ops Center (MMOC): in-house also has other clients

In contrast, the SpaceWorks CommSat has these roles:

- Customer: end user but not part of staff
- Ops team: SpaceWorks, shared among multiple clients
- Data archiver: customer-run, disconnected from SpaceWorks

High-Level ConOps Design

Your satellite might just be a free flier, tumbling through space with no active control. You might have a pulsed plasma drive and be changing orbits to head for the Moon. In all cases, you need to examine whether your satellite needs any commanding or operations needs due to its inherent limits. Some sample limits (constraints) that might need to be considered are:

- The instrument's target visibility
- Solar angle (between solar panels and sun)
- Moon contamination in FOV
- Angle to bright Earth limb
- SAA passages
- Star tracker limits with respect to the sun, moon, and bright Earth
- Contact pass availability and telemetry requirements
- Time-critical and phase-critical observation windows

- Telemetry limits, contact passes, and downlink

Pulling this together into a schema, then, you can define your ConOps as:

- Spacecraft choice, payload design, operations and flight software, navigation and control
- Commanding needs: duration, frequency, complexity, automation
- Data return needs, monitoring and trending, science needs, automation
- Team, skills, ability to resolve anomalies and faults

The first part of your concept of operations involves assessing your mission needs, the resources you can bring to bear, and the items you'll need to develop. All elements of this book feed into the ConOps—deciding what your mission goal, type, and target are define the satellite. The orbit defines its accessibility.

Next, you create assessments on how risky your mission is. Add to this the level of expertise of your people—how much experience each part of the team has in its area, and whether they have done this before.

Now assess resources—are you using established, mature software or new code? Is your development period long, or short? Is your hardware mature, or untested? Are you inheriting software or hardware from other missions, or creating from scratch?

Resources often involve compromises. Are you designing custom software or hardware to exactly fit your mission (requirements-driven), or are you using existing off-the-shelf solutions and making your mission work with their capabilities (software-driven/hardware-driven)?

Choose your physical locations. Do you have space for your MOC and SOC? Do you have support partners who are not located in the same building? Are you using an existing ground network, comm network, tracking network, or are you building your own? Draw a map of your physical facilities and how they are networked for trading data and email and communications. Include time zones, if you have more than one facility.

Next, tackle the timing. Work out how often will you need to command the spacecraft and how often will you need to downlink data. Evaluate if your solution can support this—if not, either redesign your solution or rescope your mission goals. If you are planning on automated operations, again, assess whether this exists and is mature, or still has to be developed (and whether you have the expertise).

To assess timing, list a typical set of daily (or whatever interval) items that need to be built in to the command loads (i.e., maneuvers, commands, data downlinks, types of events, calibration requests, etc.). Include an example of a typical duty cycle—the steps and intervals going from a request to an action carried out by the spacecraft.

Any implementation involves trade-offs, called "key trades." Key trades can include comparing cost versus benefit (cost/benefit), complexity/risk, speed/flexibility, development cost/operations cost (aka money now/money later), longevity/costs, longevity/complexity, and others.

Project Calliope ConOps

Here is the ConOps for my own "Project Calliope" (still awaiting launch as of first publication of this book):

- Mission type: free flier LEO satellite sampling the ionosphere constantly
- Orbit: LEO (250km) polar circular
- Tracking network: none (prediction using NORAD TLEs)
- Mission risk class: high risk, using kit parts, low heritage
- Development schedule: long (over 2 years)
- Lead level of expertise: average
- Software maturity: low—satellite: Arduino coded in C; ground: GnuRadio and custom mods; investigating use of Core Flight System (CFS)
- Number of operations locations: 1 MOC (home), 1 co-located SOC (home), open access to others

- Operations hardware: Dedicated laptop with Baofeng and larger antenna (update on original; tech has improved!)
- Commanding frequency: automated (send request for data; spacecraft returns data), no pointing needed
- Operations approach: dedicated MOC, plus amateur radio community can do downlink only
- Instrument list: ranging device, orbit capture
- Science data size: low (no images, just electric field, temperature and light scalar measurements)
- Staffing: one developer (me) plus one operator (me) plus anonymous access for downlinks (hams)
- Comm coverage: two passes of 10 minutes/day for contact pass over MOC for H&S assessment
- Commands allowed: *downlink data* and *turn off radio* (FCC requirement)
- Deorbit plan: none so far; 250 km orbit degrades within three months on their own
- Typical daily command loads: downlink of capture data and downlink of Health and Safety data

Ops Center Network

From a management point of view, there's a spacecraft, a ground station, a space-to-ground comm system, and some sort of ground network control. From a Network Operations Center (NOC) point of view, there's the antenna, telemetry, tracking, and commanding (TT&C), timing, and command. We now shift from "comms," meaning "us to spacecraft," and take a broader network view where "comms" is "from our ground link to our mission operations center." Previously we discussed how to talk to the spacecraft, right now we talk about how to talk to each other. The core "comms" view deals with TT&C.

The operational view looks at how we transport data around on the ground, handle our navigation planning, staff and run our mission control center, and deliver and analyze the data.

- Data to ops = ground station status, timing & tracking data, h&s telemetry, payload telemetry*
- Data to ground = control, pointing, spacecraft commands, voice comm views*

Organizationally, we need to set up a Mission Control Center so everyone knows what is going on and what their role is. Whether this is an actual room, or an email list among your Internet-connected distributed team, you need to be organized. To plan for handling anomalies, have an operations manual, a chain of command, and access to documentation and schematics and support material.

What makes operations complex? The primary driver is the number of uplink and downlink opportunities because you generally have to staff your operations center during those. How often you monitor your satellite, and how often you report on this, is another driver.

A major concept is duty cycle, which is the time margin available to handle routine and crisis ops before payload data is lost. If a problem arises, how critical is it that you restore functionality? For example, the Terra mission gathers 1 TB of data each day, and is able to store about two orbits worth of data. Therefore, any problem that lasts more than three hours means some data will be lost. In contrast, the older TOMS mission had a smaller data volume and could store a full day's worth of data, giving you a full day to resolve any problems before worrying about data loss. Most CubeSat missions tend to assume their data is low priority.

However, not all data is equal. HK data is important for diagnosing the status of the spacecraft. Science data may be less important, but it is often the goal of the mission. If you have a specific window for tests—deploying a specific piece a gear at a scheduled time, for example—the data during that period is more important than general monitoring data. Prioritizing when you need things to be working perfectly, and when you can accept a little downtime, will help reduce overall complexity.

One tool to aid in operations is to create scenarios for common or potential events of interest. A scenario can be *normal opera-*

tions, or *what to do when we turn on the death laser*, or *what to do if the satellite comm goes dead*. The more scenarios you predefine, the more you can ensure you have the ground tools to actually carry out that scenario if or when it occurs.

Define your scenarios by first using the orbit to set the constraints and schedule. Next, define the products needed (data down or commands up) and the actions needed. State the time you have to decide, then to generate the product, then to communicate it. Combine this to make an integrated timeline.

Once you have your scenarios, test them before you launch, on both the spacecraft bus and payload. If it's a normal operation, test it. If it's a possible anomaly or problem, fake the problem then test if you can restore the system. Done either as a walk-through or on a simulator, predefining or even automating routine activities to handle scenarios is cheaper than doing post-launch planning on ground.

Trending

A primary job of the MOC is validation of whether the satellite is working as specified in the requirements. Validation is determining whether the satellite is functioning as designed. Connected to this, verification is determining whether the mission is succeeding in its goal (usually determined by the science or technical leads). Together, they form V&V (validation and verification).

Trending is simply keeping track of the satellite over a long period of time, particularly for items that may be changing. It has several functions.

1. Predicting spacecraft performance
2. Assessing health of the spacecraft subsystems by trending
3. Tracking the use of consumables such as propellant, coolant, etc.
4. Assessing actual spacecraft performance against design goals and requirements (V&V)
5. Maintaining spacecraft parameters that may need changing or adjusting

6. Making inputs to mission planning regarding future spacecraft activities such as calibrations
7. Sometimes, validating stored commands
8. Often, maintaining flight software

Telemetry analysis, with short- and long-term trending, can find errors in all subsystems. HK trending can deduce the performance levels of the battery charge/discharge cycles, the thermal changes in the spacecraft, and the functioning of the power system. Science trending can help with instrument calibration. Navigation trending finds attitude errors, pointing problems, and propellant.

With CubeSats, it's easy enough to assess the trends just by eyeballing the data. Ideally, your HK data should be stored in a database so you can do a trending analysis at any time ("Hey, when did this battery charging problem first appear?"). This allows you to plot data over time as graphs. Formal trending tools and automated reports are standard in better operations centers.

Science Operations Center (SOC)

Science operations concerns itself with the instrument and payload data. It assumes the satellite bus is working and doesn't really care about things like *contact passes* or *the satellite is overheating*. It just wants its data. The SOC goals and requests are fed to the MOC, and the SOC receives the MOC's results. The SOC is the entity that validates the mission—the MOC can tell if you if everything is working as per specifications, but only the SOC can tell you whether the mission is meeting the purpose for which it was launched.

In terms of security, the SOC rarely has direct contact with the spacecraft, and does not send commands. Instead, they send their requested commands to the MOC, who are the only entity that can actually control the spacecraft. For a distributed project with many users, the "SOC" is the user base, and the "MOC" is the people operating the satellite. For larger missions, the SOC can act as a liaison or buffer between the MOC and data end-use customers. Because they are focused on the mission, the SOC often coordinates high-level strategic and tactical mission decisions as well.

The SOC team follows similar processes as the MOC and requires high fidelity on input (command requests in particular) to the MOC, but the SOC requires less fidelity than the MOC for real-time monitoring. A failure at the SOC won't kill the satellite, just cause less optimal performance.

In order to produce science, tasks or observations must be proposed, coordinated and scheduled, and carried out. Then, quicklook and production data must be delivered to the users or scientists. In addition, the performance of the instruments over time must be monitored, trended, and adjusted. As drivers of the mission goals (while the MOC is focusing on the mission execution), the SOC often takes a lead in creating the schedule of what to do next.

The basic concept of scheduling is simple. You take a list of goals, experiments, or targets, and make a calendar. The three main factors to consider are calculated quantities (i.e., unchangeable facts of the situation), derived quantities (soft constraints, ratings of good and bad events, and politics), and capacity constraints (such as time and telemetry).

For mission scheduling, the general calculated quantities are based on the platform. For satellites, this includes sun angle, orbital position, Earth limb angle, roll angle, and others. For ground-based units, several analogs are day/night, latitude/longitude, and elevation angle. These have no particular weight to them, but are simply physical facts about the observing situation.

From these, we create the derived quantities, which determine whether a given task or observation is feasible. This includes (for satellites) issues like allowable sun condition, occultation, thermal profile, in-SAA region, bright/dark Earth, groundstation contacts, and star tracker acquisition. Some of these are functions of the calculated quantities, and others are qualitative opinions based on the calculated quantities. Some derived quantities can be entirely political, and imposed from the outside (scientific priority, for example).

In particular, CubeSat capacity constraints include, first and foremost, time, generally defined as "you can only do one one thing at a time." Telemetry is a large concern for satellites, and other user-defined resources vary from mission to mission. Tracking all of these is the principal task of scheduling; that is, to place targets such that no constraints are violated and all resources are used to maximum efficiency without overbooking. The goal is ultimately to maximize the scientific routine.

So tools to manipulate this output not just "a schedule," but also an evaluation of its overall suitability (and stability) for the problem at hand. The interface must present all the quantities above to the user when hand-editing is required. Finally, the automatic scheduling routines (ranging from simple "good time bad time" limiting, to AI routines that make sequences) must interface with the editing options available to the user.

A good mission scheduling system can take user requests like *turn this on tomorrow* and dovetail it with other user requests (like *take a space selfie of me!*), fold them together so there are no conflicts, then actually generate the command sequences needed for delivery to the MOC. The MOC then checks that out, compares them against their own mission needs—especially contact pass times, which aren't under anybody's control—and builds the actual command strings that have to be transmitted to the satellite.

None of the SOC's criteria for a decision are necessarily the MOC's concern! The MOC must execute the SOC's provided schedule, but it is not the MOC's call as to whether it is a good, bad, safe, or risky schedule. The MOC should evaluate the SOC's proposed commands for safety—to ensure they don't overdrive the satellite or break something—but not for whether it's worth doing. Pragmatically, especially for small missions, the two groups will talk to each other, but the formal difference between the two is worth keeping organizationally to ensure decisions made are well informed:

```
Science 'not good enough' ->
      new operations modes suggested ->
              SOC builds new command sequences & macros ->
                      SOC validates ->
                             MOC validates for limits only ->
                                    MOC uploads
```

Science Monitoring

Instrument monitoring is the same as MOC's health-and-safety monitoring, only for instruments. Science assessment is a check on how well the satellite is performing its mission. A satellite can be in very poor operating condition, yet still carry out its purpose. One role of the SOC is to decide what to change if the science is *not good enough*. They can suggest new modes of operation, build new command sequences, or even discuss whether the mission needs to descope and reduce its mission goals to accommodate the true state of the working satellite.

Handling Anomalies

Anomalies are when something behaves different than expected. Most anomalies are considered a problem, which means the problem has to be either dealt with or ignored. Anomalies introduce risk. Recall that you cannot eliminate risk, only minimize, mitigate, or accept it. As the picosatellite industry matures, failures should, in theory, be reduced—but we're also always aiming for higher achievements.

The process for ground testing of hardware and for resolving in-flight anomalies are the same. There are no perfect solutions, and decisions often have to be made on the basis of partial information. It's just that on the ground, you have a lot more data and you can fix things by hand. The fundamental rule in anomaly resolution is that an anomaly, no matter how complex, has one and only one cause. The corollary is that multiple failures don't occur unless they cascade from a single root cause. Most anomalies can be resolved without changing operating modes. If anomalies indicates danger to the spacecraft's health, command it into a safe haven. If in doubt, abort the current operations and safe it.

Apollo 13 Anomaly Tracing

The Apollo 13 mission to the Moon had to be aborted when the spacecraft developed a problem. This oft-cited case (and resulting movie) is a good look at anomaly tracing and root cause analysis (RCA).

Initial event sequence
: Crew hears a *loud bang*, their spacecraft begins moving erratically, and the voltage levels are fluctuating.

Bad analysis
: Spacecraft suffered a simultaneous meteorite impact, spontaneous power-on of thrusters due to bad software, and fuel cell problems all occurring at the same time? No!

New data
: Loss of air in an oxygen tank. Also, fire.

New data 2
: Debris (bits of metal and such) seen floating outside the spacecraft.

New data 3
: Other oxygen tanks began to leak.

We build an RCA chain. A fire can't just start up, it needs a cause. Likewise, a fuel cell losing charge won't start a fire; instead, something must cause the fuel cell to lose charge. Debris floating outside could indicate an interior explosion or a meteorite event. The thrusters firing erratically is an electrical problem. An explosion can cause an electrical problem, but then we have to figure out what can cause an explosion, and the answer is either circular (*an electrical problem in an oxygen tank*) or external (*something hit us*). A short-circuit in wiring can cause the electrical problem, but then we need to figure out what caused the short circuit (either *chance* or *manufacturer's defect* being likely possibilities). The other oxygen tanks would begin to leak in any scenario involving an explosion in one oxygen tank. Also, the other fuel cells could fail if there is an explosion or leak in the oxygen tanks.

So our chain is either "a defect in the wires caused an electrical short-circuit, that led to fire, other oxygen tank trouble (which then causes more fuel cell trouble), random thruster firing, and an explosion of debris" or "an explosion caused an electrical short-circuit as well as debris, which led to fire, other oxygen tank trouble (that then causes more fuel cell trouble), and random thruster firing."

This analysis is very helpful, because it sets the key order—first there was a single cause (either meteorite or defect in wiring), *then* the other events follow. At this point, we have a single root anomaly. We don't yet need to find whether it was a defect or meteorite, because the sequence tells us the main problem was *something* caused a short-circuit, that led to the other problems. From this, we can begin to assess the likely level of damage, which systems we can still trust, and what the best recovery is!

Availability analysis looks at mean time between failures (MTBF) and mean time to restore (MTTR) where availability is A:

$A = MTBF/(MTBF+MTTR)$

(e.g., MTBF = 400 hrs and MTTR = 2 hrs → A = 0.995). In other words, over the course of 1 year such a system is unavailable 16.7 times for a total of 43.8 hrs! Fun fact—a survey of 13 spacecraft over 2 years indicates anomalies occur about every 7 months of spacecraft operations. Statistically, then, even the most robust spacecraft should expect at least one anomaly. For CubeSats, built fast and cheap, expect more.

The core of resolving anomalies is root cause analysis (RCA). You can get very deep into formal techniques. One is formal failure modes, effects, and criticality analysis (FMECA). In this process, before launch you define conditions and constraints for each component, define what constitutes failure, and assign probabilities and severities. From this you can predict many anomalies before they happen. In CubeSats, most people do not have the staff depth nor experience to do this in detail. So we go to diagnosing anomalies as they happen.

Now that you've saved...

1. Get all the data. All.
2. Establish accurate timing and sequence of telemetry events.

3. List possible causes.
4. Analyze data and find the cause (look for single points of failure, construct a failure tree).
5. Determine the corrective action.
6. Carry it out (take your time, verify each step, do no harm).

A sample failure tree for "magnetometer not recording" could be:

- Not getting voltage
- Component damaged
- Connection to CPU down
- Software not reading it

Multiple failure trees can help drill down to a single cause. You need to determine if additional analyses or spacecraft HK data are needed to define the root cause. Once you determine the root cause, then (and only then) are you ready to start looking at ways to solve the anomaly and restore normal operations. Normal operations—in fact, boring operations—is the goal of any well run Mission Control.

Root Cause Analysis Example

Taken directly from Six Sigma (*http://bit.ly/1E5EKJA*).

Problem statement: You are on your way home from work and your car stops in the middle of the road.

1. Why did your car stop?

   – Because it ran out of gas.
2. Why did it run out of gas?

   – Because I didn't buy any gas on my way to work.
3. Why didn't you buy any gas this morning?

   – Because I didn't have any money.
4. Why didn't you have any money?

   – Because I lost it all last night in a poker game.

5. Why did you lose your money in last night's poker game?

   – Because I'm not very good at bluffing when I don't have a good hand.

Ergo, root cause analysis reveals that your car stopped because you are not very good at bluffing in poker. So stop playing poker, or the problem will occur again.

A/Resources

We reference several useful books, and also provide additional links to specific projects. We encourage you to look at past-flown and proposed mission architectures, in the process of evaluating how you will do communications. The field of amateur satellite communications is rapidly changing and by the time you think you might have a problem to solve, someone else has probably solved it for you!

- Android ham tools: Wolphi—redirects to wolphi.com (*http://www.wolphi.com/android-apps*), Amateur Radio Android Apps (*http://bit.ly/1E5F8re*)
- ArduSat Learning Center (*http://bit.ly/1Gp83wM*) and ArduSat Freetronics (*http://bit.ly/1E5FWMQ*)
- American Radio Relay League (ARRL) (*http://arrl.org*) and the International Amateur Radio Union (IARU) (*http://iaru.org*)
- Useful CubeSat Kit Links (*http://bit.ly/1E5G7b4*)
- DIY Rocket Science (*http://stephenmurphey.com*)
- Don't buy used (radio advice): AR15.com (*http://bit.ly/1E5GjXH*)
- Golden screwdriver: both good or risky (like overclocking) (*http://bit.ly/1E5Gpi3*)
- IEEE Xplore Abstract - Communication and data handling system for BRICsat satellite (*http://bit.ly/1E5Gq5P*)
- Interface Android device to radio (*http://bit.ly/1E5Gu5z*), Toshiba Thrive Forums (*http://bit.ly/1E5GzpG*), Wolphi-Link Interface (*http://www.wolphi.com/interface*)
- Power-CySat-Iowa State University (*http://bit.ly/1E5GDWH*)
- Project Calliope (*http://projectcalliope.com*)

- SatNOGS • Hackaday Projects (*http://bit.ly/1E5GQsM*)
- $50SAT - Eagle2 (*http://www.50dollarsat.info/*) and their "$50SAT PocketQube Amateur Radio Challenge" at AMSAT-UK (*http://bit.ly/1E5GUc2*)

Afterword

DIY Space mixes the known and the unknown, requires a good idea but moves past it, and then hurls that realized idea into one of the harshest environments ever to see if it survives. Over the past three years, we have shifted from asking whether amateur space is possible and instead moved to the fundamental question of what will you do and why are you doing it. We're at a point where the tech is plentiful and information is exchanged freely.

The truly tricky part of DIY Space isn't the space part. It's that, by the time you read anything, someone else has already pushed things further, built something more clever, or made something hard become simple. DIY culture is an awesome mix of sharing things that work and fighting each other to prove the impossible. Everyone helps newcomers enter the DIY world, riffs off each other, and pushes each other to excel, all at the same time.

DIY is more than ideas, but certainly a unique idea is a treasure. In this, there's a self-promoting tendency of DIYers each trying to claim *first* for their project. I say let us—every new incremental step moves us all forward, and a little egotism can be a good motivator in DIY culture. Just be careful to maintain balance in understanding that your new idea is built on a shared culture, that even mavericks are boosted by the efforts of others. Do that, and you can score any new *firsts* you can dream up. I'm waiting for *the First Mauve Satellite Funded by Cookie Sales*, myself.

Or, simply enjoy building something yourself, with or without novelty. If it's new for you, that's new enough. Just the ISS NanoRacks deployment program pushed out 33 new CubeSats in a span of a few weeks—and I guarantee not all of them are *ideas never flown by humanity*. But equally guaranteed is that each is new for the DIYers and universities and small groups that built the satellites.

What distinguishes DIYers from spectators is we move past the idea into actually building and executing our mad schemes. For space, it's worth understanding what Chris Scolese (the NASA/GSFC Center Director) meant by stating that integration and test is the critical path for success. This is both the key and missing piece for DIY Space to become more mature. Integration at its core is just taking stuff that works separately, and getting them to play nice with each other. Testing is either the coolest or most boring part of DIY. Cool, in that you get to try to break your new stuff; boring, in that you have to do it over and over. Embrace both, so when your project hits space, it succeeds.

Participating in DIY Space doesn't require everything to push the envelope. Just mixing known solutions yields coolness. The "ISSAbove" project takes an ordinary Raspberry Pi, some known software solutions for predicting orbits using data fetched from the Web, and turns it into DIY coolness by making a box that lights up when the ISS is above your house. Folks are taking ground-learned solutions for constellations and swarms of robots and 'coptors and starting to apply them to satellites. Other folks are simply working to make it easier to communicate and operate space stuff. DIY operates brilliantly when taking cross-disciplinary ideas or simply in combining known stuff in a way no one combined it before.

Don't worry if this makes space seem ordinary. Space is still a really hard target to hit. Space is a hostile environment, one of the big three. Space, ocean, and inside the human body all share weird mixes of pressure problems and materials behaving oddly, combined with sheer inaccessibility and the fact you can't really fix mistakes once they happen. The tech is there, but the key to space is realizing you have to execute your project really, really well because you have only one shot. There's no post-launch tweaking or second chances.

There are so many areas where DIY Space is really active. Obviously, actually building and hurling DIY satellites into the void is hot and gets a lot of press. It's not just tech demos, either. Folks are starting to tackle science problems using pico- and nanosatellites, going to orbit to gather ionospheric data or remove orbital debris or spot asteroids. There's a LunarCubes movement tackling the need for nanosat propulsion to get past low Earth orbit and visit the moon, asteroids, and planets. And of course PocketQubes and sprite concepts are actually flying and proving that we can send satellites that are smaller than ever yet do 100% of what's needed.

DIY Space also isn't just space. Efforts at building up a stable (often open source) infrastructure in terms of hardware, radio work, and ground stations is quieter but growing just as much. Space awareness—projects that listen to space, look at space data, or connect us on the ground with what's going up outside our Earth in a meaningful way—is starting to heat up as DIYers use clever methods to tap into existing space-based data. As the DIY culture builds, we're seeing STEAM efforts that add *Art* to the usual STEM approach, which draws more people into what we do.

We use 3D printed CubeSat *shells* for test builds, high-altitude balloon payloads, and sensor work. Once you have a cheap, reliable form factor, it can be used for anything. A satellite is just a robot that orbits; a sensor box is just a nonmoving robot, and a drone is just a free-flying nonorbit satellite:

In short, DIY has it all as a mix of proven and new, first tries and sustained efforts, tech and science and art, all powered by people and going to space. It's a good time to look past Earth, and everyone is invited along for the ride.

About the Author

Alex "Sandy" Antunes is an astrophysicist who turned to science writing upon realizing his desire to understand the universe doesn't mean you have to be the one to discover everything personally. There's a lot of excellent science out there, and Sandy enjoys bringing it to the world's attention. Sandy recently achieved a professorship at Capitol Technology University's Astronautical Engineering department, which he credits to his NASA work, his solo build of the Project Calliope picosatellite, and his writing for Science 2.0 and via Make: and O'Reilly Media.

DIY Comms and Amateur Space is the fourth book in the four-book *DIY Satellite* series. The first, *DIY Satellite Platforms*, covers the basics in designing and building your own satellite. The second, *Surviving Orbit the DIY Way*, is about the environment of space and how to create a survivable design. The third, *DIY Instruments for Amateur Space*, covers the instruments and science you can achieve in space.

8/4/15

CPSIA information can be obtained at www.ICGtesting.com
Printed in the USA
BVOW08s2354010615

402764BV00002B/3/P